William Ludlow Coleman

A History of Yellow Fever

William Ludlow Coleman

A History of Yellow Fever

ISBN/EAN: 9783337253417

Printed in Europe, USA, Canada, Australia, Japan

Cover: Foto ©berggeist007 / pixelio.de

More available books at **www.hansebooks.com**

A HISTORY

OF

YELLOW FEVER

Indisputable Facts
Pertaining to Its

ORIGIN AND CAUSE

❧ ❧ ❧ ❧

And its present artificially acquired habitat, with reasons going
to show the possibility of its complete extinction from
the globe, its nature, anatomical characteristics,
symptoms, course and treatment with
an addendum on its twin sister

DENGUE

Containing a parallel table of the most prominent symptoms
of each disease.

❧ ❧ ❧ ❧

BY

W. L. COLEMAN, M. D.

Houston, Texas.

DEDICATION.

This book is affectionately dedicated to the Medical Profession, in which I have been a hard and earnest worker for over forty years, with the hope that through them it will be the means of accomplishing much good for my fellow men.

THE AUTHOR,

PUBLISHER'S STATEMENT.

The Author's manuscript was prepared during the year 1897 and has been carefully and painstakingly revised in the office of publication, to comply with the political conditions now existing. It is believed that the Doctor's predictions are correct and that we shall never have another widespread epidemic of yellow fever, particularly if his suggestions for the prophylactic treatment herein expressed are faithfully carried out. And if cases do occur we believe that the author and his many co-workers will be able to demonstrate that the disease can be more successfully combatted along alkaloidal lines than it has ever been before. We therefore heartily commend the little book to all who would know the truth of the past, the present and the probable future of yellow fever.

THE CLINIC PUBLISHING COMPANY.

Chicago, July, 1898.

PREFACE.

While the launching of this frail little craft,
pilotless as it were and from an unknown
builder, upon the literary ocean of medicine,
and especially upon that part of that turbulent
sea the surface of which has been so frequently
agitated by the fiercest and most bitter wordy
combats and violent controversial storms concerning its subject, may be regarded as an act
of foolhardy temerity; yet although it is the
author's first venture in this particular department he sends it forth fearlessly, utterly ignor-
ing and indifferent not only to the "*imprimatur*"
of the celebrated French Academy of Sciences,
so highly prized and desired by Dr. Audouard,
but also to the endorsement of all the so-called
scientific societies of the world, fully convinced
and well satisfied that it will be eagerly sought
for and read by that large and rapidly increasing body of intelligent, progressive physicians,
"The Burggrævian Dosimetrists of Europe,"
and "The Abbott Alkalometrists of America,"
who, being dominated solely by that rare
quality of mind, common sense, "prove all
things" and are in no way governed by the old
rule of "Magister dixit."

I do not mean the least disrespect or dispar-

agement of them when I say I ignore the learned
scientific bodies of the world and do not care
the snap of my little finger whether or not they
read a line of my unpretentious little collection
of irrefutable truths concerning the subject of
my work. They ignored Dr. Audouard's valuable brochure because it did not bear the seal of
the French Academy, although he was a special
representative of that society on the official
commission appointed by the French government in 1821-23 to investigate the yellow fever
epidemics of Spain; while mine is not only
entirely destitute of a seal from any kind of a
society, but it is the production of one who
determined forty-five years ago, when a student,
to discover all possible hidden truths of this
disease about which the most brilliant and
gigantic minds in the profession had been
wrangling for nearly a century without coming
to a clear understanding about any of them.

And now, without a scintilla of egotism or
spirit of boasting, I present as a result of my
labors the chief of those old points of contro-
versy cleared of all mystery and established
upon the firm basis of immutable truth; and
doing this in a spirit of perfect confidence born
of this result of my labors, I demand rather
than solicit, as authors usually do, the attention
and investigation of the rank and file of the profession—that great body of hard workers who
always honestly desire and earnestly seek for

the truth in such questions, well knowing that
correct diagnosis and successful treatment cannot be attained without a perfect knowledge of
the origin, cause and nature of disease. Hence
as it is to their interest, I expect they will be
my principal readers and I trust they may be
greatly benefited by my labors; but whether or
not I am mistaken and am premature in my prediction that the scourge has so nearly run its
course, and that it will never produce a general
and destructive epidemic in this country again;
yet it is reasonable to expect that on account of
the present strained state of affairs between
Spain, Cuba and our country, there is great
danger of a more wide-spread epidemic of yellow
fever this summer throughout the length and
breadth of our land than ever has occurred or
ever will again occur; and in that event, there will
be a greater number of this class of physicians,
called upon to treat this disease than ever before
in its whole history. Naturally our soldiers
will as a rule be non-immunes, and I fear, alas,
that more of them will be cut down by this terrible disease than by Spanish bayonets and
bullets.

It is under just such circumstances that I
hope this little brochure will be of great help and
benefit; for while I have not given the profession
a treatment that will always cure the disease,
none having been discovered, yet they will find
in this work advice, directions and warnings for

preventing or modifying it that cannot be found elsewhere and which, if heeded and carefully followed, will save many valuable lives.

I have no grounds for egotism or boasting, as nearly all the facts presented were obtained by eminent men all along the line down to the present; but I have been enabled by the evidence of subsequent historical events to use these facts in deducing and establishing truths which they could not see for want of this later evidence, thus proving Audouard's theory to be no longer theory but a stern, irrefutable truth.

W. L. C.

Houston, Texas, U. S. A.

INTRODUCTION.

"To all facts there are laws;
The effect has its cause, and I mount to the cause."
—*Lucile*.

The world is too busy to form its opinion on a question at first hand; hence important his-torical facts and truths are seldom recognized or understood by people of the age in which they occur, for some are at times suppressed before suitable knowledge is obtained of them, some for political and some for commercial reasons and yet others from bigotry; so it requires time for these different objections to pass away, as well as for the corroborative evidence of subsequent events to establish them beyond question as facts and truths. In the course of this paper I will mention a marked instance of the suppression of truth for each of the aforesaid reasons, and thus show that even medical men are often blinded to the truth by their prejudices and passions. We of all men should be liberal-minded, free from bias and bigotry and incapable of being swayed by polit-ical, commercial, religious or other influences that tend to obscure or suppress truth.

I have been intending for some time past to undertake the task of collecting and arranging for publication the records of facts and truths that I have obtained by personal experience, observation and investigation of yellow fever

during more than forty years, together with
everything else of interest and scientific value
that I have obtained from reliable sources which
would help to make a correct history of the disease. But procrastination has been the bane of
my life; and I fear I have delayed the work till
it is too late to accomplish it in a manner that
will render it interesting and profitable to the
present and future generations of medical men.
My feeble condition and rapidly failing strength
warn me that but little time is left in which to
work. Would that some younger and more
fitting hand and brain had the task!

  In the beginning of my medical studies a
fascination for the subject of yellow fever took
possession of my mind and I have never been
able to wholly shake it off. During many weary
years of study I wandered blindly through the
mazy labyrinth of mysteries surrounding it,
seeking light and truth but finding none. At
the point of despair I was led from that state of
uncertainty by a ray of light emanating from a
gem of truth seventy-five years old which reached
my dull brain by penetrating through the prejudice and passion that had resisted its entrance
for forty years; and I set to work to bring order
out of the chaos of facts and truths which had
been gathered by myself and others during the
past history of this horrible scourge.

  The two greatest barriers to the reception of
truth, prejudice and passion, having been re-

moved, it became easy to accept and understand the plain solution of the etiological "x" in the problem stated by Dr. Audouard seventy-five years before and to see its complete verification by every event in the subsequent history of every thing connected with the subject. It was about this vexed point that some of the fiercest and most violent controversial storms raged that ever disturbed the medical mind; while the question of its contagion, nature and connection with malarial fever came in for their share in the controversies which, with each recurring epidemic, were indulged in by the giants of the profession from the days of the great Benjamin Rush down to those of Doctors La Roche and Warren Stone.

Surveying those wordy contests, which were conducted with so much heat and passion of argument, by the light derived from the knowledge of its true origin, shows them to have been supremely ridiculous; for it reveals the ignorance of both parties concerning the question under discussion in which both claimed to be, and honestly thought they were, exactly right. It was always a mystery to me why they so stubbornly adhered to their respective opinions, for they were men eminent in the profession who readily yielded to reason upon other subjects. The mystery has been cleared up by the new light, the knowledge of which however lay hidden for nearly seventy-five years; and in the course of my discussion of that point it will readily be shown that both parties were right at that time, however paradoxical this assertion now seems.

For more than two centuries this unique disease was an unsolved mystery standing solitary and alone, as has been said of the great Napoleon, "grand, gloomy and peculiar, wrapped in the solitude (mystery) of his own originality." Claiming no land as its birth place (and it never has originated *de novo*) it is absolutely and wholly a filth disease, a specific animal poison of peculiar filth, born or generated upon the high seas and under peculiar conditions (which have ceased to exist), as a result of man's violation of Nature's immutable laws and the inexorable fiat of their Author that the violation of one natural law brings its own punishment by setting into active operation another law governing the changed condition. It can truly be said to be "the pestilence that walketh in darkness, the destruction that wasteth at noon day;" and after years of study and investigation of this scourge I can come to no other conclusion but that it is a Nemesis, an agent of retributive justice in the hands of Him who hath said "Vengeance is mine, I will repay." But since the cruel traffic to which it owes its origin has long since ceased, and the institution of slavery, of which J. K. Ingram said (Encyclopedia Britannica—"Slavery"): "It was politically as

well as morally a monstrous aberration, and never
produced anything but evil," has been abolished
by all the nations of the earth, I have believed that
this instrument of punishment would also be removed. After finishing my-work in Memphis in
1879 I predicted that there would never be another serious and widespread epidemic of this
disease in this country; for I had the presump-
tion to think that the South had received its full
meed of punishment for the part it took in that
great sin, that is, if we reckoned in the estimate,
the horrors of the four years of civil war as
legitimate results; while the North received its
portion from 1693, the date of its first appearance in this country, down to 1822, since when
its visits to that region have been few and far
between and confined to small areas.

In the following history of yellow fever I hope
to be able to establish as an irrefutable truth
Dr. Audouard's theory of its origin, and thereby
refute and show the littleness of the damaging
epithet of Prof. Hirsch who, in describing it in
his treatise on "Geographical and Historical
Pathology," characterized it as "eine der
abenteuerlichsten hypothesis," a Quixotic idea
unworthy of credence. And I think it appropriate to remark just here, if I succeed in this
object, that the Royal Medico-Chirurgical Society of London, in whose library a copy of
Audouard's book lay with uncut leaves for sixty
years, certainly owes it to his memory and to

the medical profession in general to resurrect
and republish his work as a partial reparation
for their neglect and quasi suppression, for more
than half a century, of a book containing a truth
as grand and useful as any other discovered
during this wonderful nineteenth century.

Before going into the chronological history of
yellow fever, I deem it necessary for a clear
understanding of my design to give my views
and final conclusions in regard to this disease.

1. Yellow fever is not native to any continent of the old or new world, and never *per se*
originated *de novo* upon any spot of land upon
the globe.

2. Yellow fever is a specific disease peculiar
to itself; and while there are several apparently
analogous diseases, yet it stands separate and
distinct from all others. It is always identical
in every climate where it prevails and is not a
grade or type of malaria or other zymotic fevers.

3. Yellow fever is pre-eminently a filth disease, caused by a specific infection or animal
poison generated from a peculiar filth and under peculiar conditions, and possessing a germ
capable of transportation and reproduction.

4. Yellow fever is never contagious but is
highly infectious; and its germs, while unlimited
as to quantity, are limited to two stages of existence in their power of reproduction.

5. When this germ resulting from that specific poison is introduced into a locality where

where the unsanitary conditions and the meteorological stage of the atmosphere are suitable for
its existence and propagation, the second stage
is ushered in by the production of the true
pathogenic microbe of yellow fever in such
quantities and numbers as to fill the atmosphere
of that region as far as the suitable conditions
extend.

6. When a sufficient portion of this specific
poison once enters the human organism and
produces its characteristic phenomena, its vitality
ceases and it becomes so innocuous that a healthy
person may drink the blood or be inoculated
with the black vomit or other secretions and
excretions of a fatal case of yellow fever with
impunity, thus verifying the teaching of thousands of observations and centuries of experience that personal contagion in this disease is
an impossibility.

7. For the same reason what are known as
hibernating germs or microbes, or those which
have failed to find a field in which to propagate
and at the end of the epidemic are so favorably situated as to have their vitality and virulence pre-
served from destruction by the cold of the succeeding winter, can only produce sporadic cases
the following season—never an epidemic. I

it is plain that if the preceding proposition is true, this necessarily follows as a natural and logical sequence.

8. And finally: the peculiar filth which originally gave rise to the specific infection of yellow fever was for nearly two centuries deposited in a number of the harbors of our Atlantic seaboard and in the alluvia of the Mississippi at New Orleans, and this in quantities sufficient to feebly naturalize or render it indigenous to these localities. But the rushing, rolling, raging billows and the fierce, passionate, pitching, plunging tides of the Atlantic have long since washed every trace of it from the mud of every American port from Cape Cod to the Cape of Florida; while the massive volume of water annually rolling down the Mississippi has cleared it out of the port of New Orleans where it has been more frequently claimed to be indigenous than at any other point on our continent; hence the value of strict and perfect coast-line quarantine.

PART I.

ORIGIN AND CAUSE.

"Felix qui rerum potuit cognoscere causas."—Virgil.

Perhaps no disease in the annals of medicine exhibits so many unique and remarkable facts in pathology or has elicited as many fierce controversies and differences of opinion as has yellow fever. These hotly debated controversies as to its nature and origin, more especially the latter, have been conducted and engaged in by some of the most brilliant medical minds of America from the days of Rush, Physick and Currie down to those of Laroche, Dickson, Campbell, Warren Stone and others.

Until 1766 there seems to have been no disagreement in the profession as to its origin, being always regarded as an imported disease; and by common consent its home was located exclusively in the West Indies in general, and in Havana in particular. But the earliest historians of those islands ascribe its origin to the East, and for many years from its first appearance there it was known as the *"mal de Siam."* Subsequent history proves those writers to have been perfectly correct as to the direction whence this terrible scourge came; but they located its

ce too far east, for neither Asia nor
in be truthfully charged directly with
l.

opinion, formed from a careful study
ew of the whole history of yellow
ere could be no more appropriate place
riginal home on this little globe than
rious "middle passage" with all its
t atrocious horrors.
facts" about which those giants in the
profession wrangled and contended so
during the last quarter of the eighteenth
ire now proven to be errors, at least
hem; while a few, which were facts in
ve ceased to exist from removal of their
us showing that facts as we see and
end them are not always final truths.
s immutable and eternal," yet often the
upposed) of today is the error of
v. At the proper time I hope to be
1ake good my promise to show clearly
unprejudiced mind that both parties in
erce controversies were right at that
would not be so now.
fever in its history has been so inti-
connected with the old African slave
its rise and decline, both as to time and
it only at the ports of debarkation in.
ern but also in the ports of the Eastern
ere where the disease prevailed so vio-
1enever one of these vile, filthy ships

cast anchor in the harbor on the retu
that the history of the one necessarily i
that of the other. This singular but c
association of the two strangely seems
been entirely overlooked by all the early
as the later writers upon the disease. I
anyone of them, so far as I know, not
made special mention of the decline and
pearance of yellow fever from our most
tant northern ports as soon as the le
slave trade was abolished, where it h
vailed frequently and virulently for more
century after its first appearance in Bc
1693. These places, wherever the dise
occurred, had also been ports of debarka
the slave ships during the long period w
trade was recognized as legitimate and
on principally by England.

For nearly two centuries the foul, fern
and putrefying dysenteric discharges, alo
other filth from the poor negroes which a
lated in the holds and poisoned the b
hundreds of slave ships, had been pum
to mix with the mud of those harbor:
quantity of this peculiar filth finally
sufficient to give rise to yellow fever, thus
ing it feebly indigenous and endemic a

where with the same traffic which had now
fallen into the hands of contrabands and was
carried on by stealth; and as those contraband
ships sought ports only where their cargoes
were wanted, the disease accompanied them to
our southern ports, the gulf ports of Mexico and
the South Atlantic ports of Brazil.

This plain and indisputable historical fact
seems to have attracted but little notice and
elicited but little comment in the past from
the historians of yellow fever. This oversight
emphasizes my first statement that such facts
are seldom recognized and understood at the
time of their occurrence; and sufficient time must
elapse for the passions and prejudices against
them to subside and pass away before they can
be looked at in the right light and established
as truths. But the greatest oversight of all was
in regard to the fearful epidemics which ravaged
the ports of Spain during the first quarter of this
century, causing the death of over one hundred
and forty thousand persons in the one year of
1804. So, in my opinion, any period prior to
the great epidemics of 1878-79 which occurred
in the Mississippi Valley from New Orleans to
Memphis was too early in which to write the
correct history of yellow fever, a disease which
I confidently believe to be now making the final
chapter in its history.

The first authentic appearance of yellow fever
was in 1647, shortly after the first arrivals of

slave ships in the West Indies, a terrible
epidemic occurring at Bridgetown, Barbadoes,
an English possession and settlement. It was
spoken of at that time as "a new disease" and
as "an absolute plague, very infectious and
destructive." But the Jesuit narratives of the
period of Columbus describe a similar fever of a
malignant type, attended with yellowness of
skin and very fatal to the newly arrived Euro-
peans, prevailing epidemically in St. Domingo
and other islands in 1494, 1514, 1568, and in
Central America in 1596, and among the Indians
of New England in 1618. Some writers on the
subject claim these epidemics to have been
genuine yellow fever; but there being no account
of these fevers, except by non-professionals, the
large majority of reliable historians oppose this
view and have agreed upon 1647 as the date of
the first authentic record. It was not till 1691,
after the arrival at Martinique of the Oriflamme
and two other vessels loaded with French colo-
nists fleeing from Siam on account of an insur-
rection, that it was first called the "*mal de
Siam*" and was so designated-for many years
thereafter.

The true history of these three vessels prior
to that trip would be a very interesting as well
as useful subject for investigation; for it would
be strong corroborative evidence in favor of the
truth of Audouard's theory if the records would
show that they had been old slavers. There are

good grounds for the belief that such was the fact, but I am unable at this writing to find the record of my authority for this statement.

Its first appearances in the United States were at Boston in 1693, in Philadelphia and Charleston, South Carolina, in 1699, and in New York in 1702; and from the first named date, 1693, down to 1879, a period of 185 years, there were 115 epidemic years in which the disease visited 338 different cities and towns in this country, many of them repeatedly and some almost every year of its appearance. New York, Philadelphia and Charleston, South Carolina, were visited most frequently during the period of the so-called legitimate slave trade; they being the greatest centers of population were the principal ports of debarkation for the miserable captives. Its most frequent and regular visits were during the contraband slave trade from 1808, the date of the abolition of the legitimate traffic, down to 1860, a period of fifty-two years. During this period it failed to appear in two years only, viz., 1813 and 1836. And the record shows the significant fact that the first named cities, New York and Philadelphia, were practically exempt during that period and that the disease appeared almost exclusively at our

undeniable facts; for they are *prima facie* evidence
of such a close and constant connection between
the slave trade and yellow fever as to force us
to the unavoidable conclusion that the relation
was one of cause and effect.

All the facts presented by the history of the
disease since 1860 go to strengthen and confirm
the above conclusion; and it is plain to be seen
by any one who has closely watched the course
of yellow fever for the last forty years that it has
disappeared from the harbors of the United
States as an indigenous disease; and the natural
inference is that this is the result of the cessation of the annual addition of the peculiar filth
from the holds of African slave ships to the mud
of those harbors. If I thought it would be
deemed necessary or demanded by the profession I could easily give the names of all the
places visited, with date and mortality of each;
but it would require fifteen or twenty pages of
names, dates and totals, which would be dry
reading and useless except for reference. It is
on record and can be found in several histories
of yellow fever and I therefore will omit it.

Yellow fever visited Cadiz, Spain, in 1705,
the first authentic record of its appearance in
Europe; and yet the venerable Dowell, who was
my personal friend and whose memory I revere,
says in his immense volume on yellow fever
(which by the way contains many absurd and
unwarranted assertions and contradictions): "It

was undoubtedly introduced from Africa to
America. [No reason or authority given for this.]
That it existed in Africa, eastern Asia and
southern Europe long before the establishment
of the Greek and Roman empires is generally
well established by Hertardo, even running
back a thousand years before Christ; that it has
become endemic along the coasts of Africa, both
east and west, as well as in the West Indies and
northern coast of South America, and that in
all these districts it has its epidemic years and
its years of nearly entire exemption, is also well
known."

But Dowler, who is considered better authority,
says: "The slightest notice of yellow fever is
nowhere found among ancient writers, although
they have not failed to record incidently or
directly the time, place and progress of numerous
epidemics with more or less particularity, so
that their characteristics may now, after the
lapse of centuries, be ascertained."

Yellow fever does not appear to have been
noticed until after the discovery of America by
Columbus. Had it prevailed in ancient times,
its prominent features, so very remarkable at
least in its advanced stages, would doubtless
have been recorded. From 1705 to 1808 it
visited Cadiz nine times, but after that it broke
out and raged like a prairie fire all along the
coast of Spain, appearing frequently at all the
important ports till 1823. Why was Spain

alone of all Europe thus scourged? Because she was the first nation to. inaugurate the African slave trade and the last to abandon it; because the contraband ships were nearly all owned by her citizens and the illicit trade was thus virtually carried on by her tacit permission, which gave easy entrance into her ports for these vessels and which was denied them at the ports of all the other nations; and so they came on the round voyage from Africa to the West Indies or the Brazils and back laden with the products of those countries—cotton, sugar, tobacco, etc.; and although they generally came with "clean bills of health" from Havana or Rio, yet their holds were reeking with the peculiar filth that gave rise to this Nemesis-like scourge. It is estimated that the population of Spain was diminished one million by this disease alone; the official report of deaths from yellow fever for one year amounted to one hundred and twenty-four thousand.

The disease spread to a greater extent in the interior than it ever did in any other country, visiting twenty-five cities and towns in the year 1804. In some of these the number of persons affected amounted to 1 in 2.78 of the population; and the average proportion of deaths to the number afflicted was 1 in 3.087, the extremes being 1 in 1.3 and 1 in 6.42. The extreme virulence of the disease and the great mortality was undoubtedly due to the fact of the specific

poison coming directly from the holds of those
slave vessels which were never properly cleaned;
and thus Spain received her full meed of punishment for the part she took in this great crime
against humanity.

The oversight of yellow-fever historians and
of the profession generally in the matter of the
Spanish epidemics forcibly illustrates how super-
ficial we are at times in our observations and
investigations of a subject; for although
Audouard pointed out the path of truth in 1825
and the facts in the case are now historical and
as plain as day-light, yet no attention has been
paid them and little or no importance attached
to them. The facts to which I refer are all
included in Audouard's discovery made while
officially investigating the epidemic in 1821 and
1823, viz.: The cause of the prevalence and final
disappearance of yellow fever from the coast of
Spain. And I add, why were these particular
ports thus scourged while the rest of Europe
escaped?

It is true that Dr. Audouard's report of his
discovery and his essays containing his theory
were ignored and, to a certain extent, suppressed
by the French Academy of Science by the with-
holding of their "*imprimatur*" for political reasons, fearing to offend England; and so his book
was shelved in the libraries of the scientific
societies where it lay with uncut leaves and
covered with dust for more than half a century,

positive evidence that not a single officer or
fellow of those learned bodies ever read a line
of it. But there was still another reason why
Audouard's theory attracted so little attention in
Europe; and that was that those terrible epidemics of yellow fever in Spain were over,
having ceased with the removal of the cause;
and thus the danger of future epidemics being
also removed, all interest in the disease ceased
and no future investigations were made because
they contended that it was a disease wholly of
the Western Hemisphere and that it always had
been imported whenever it had prevailed in the
Eastern, never having originated at any time or
place on that side of the globe. But the first
historical reference to yellow fever ascribes its
origin to the East; and even down to the present
time there is a vague idea in the minds of medical men that, though it had become domes-
ticated and indigenous in the West Indies, it
originally came from Asia, the home of so many
dreadful plagues.

A calm and impartial review of the history of
the disease reveals the fact that there is no
ground to sustain either opinion; and it now
devolves upon me to state the facts and reasons
for my opinion that it never originated upon
land but that it was born upon the sea. I do
not mean by this that it was born or generated
originally in the sea water itself, though it seems
to have become an aquatic bug in its artificially

acquired habitat. The only spot on earth where
it can now be truly said to be indigenous and capable of constant reproduction, is in the foul and
offensive waters of the low land-locked, almost
tideless and stagnant Bay of Havana, with its
one single, narrow outlet. We will discuss this
point more fully later on; but let us now take a
map of the world and paint a yellow dot where-
ever yellow fever has been known to prevail in
the past, beginning where it first appeared on
our Atlantic seaboard and has since disappeared,
using the lightest shade and deepening the color
as we trace it in close companionship with the
slave traffic, declining southward until we reach
our gulf ports and the West India Islands,
where the deepest shade must be used to indi-
cate that there it still exists. A few Mexican
and South American ports must also be colored;
but crossing to the Eastern Hemisphere, Europe
requires no paint except at certain ports in
Spain already mentioned. Africa requires only
two or three small dots on its western coast
where slave ships trafficked and obtained their
cargoes, while Asia, long supposed to be its
original home, takes no paint at all. It has
often been asked why the ports on our Pacific
seaboard were never visited by the scourge, and
the only rational answer is that slave ships never
traded on that coast. The only points on the
whole line of the Pacific seaboard where the
yellow fever ever did obtain a foothold were Cal·

lao and Lima, Peru, in 1853; and its outbreak at these points is strong corroborative proof of the truth of Audouard's theory. It was universally ascribed to the arrival of filthy ship-loads of Chinese coolies who were brought over by contract, as the Africans had been, in badly equipped and over-crowded ships and most of them suffering terribly with dysentery. And as Audouard reasoned in the case of the Africans that the specific poison which caused yellow fever having come from the negro body could not poison the negro again, not being auto-inoculable, therefore he was exempt from the disease, so it was found that the Chinese in Peru were as much immune from yellow fever as the negroes on the Atlantic Coast.

The intimate association and close connection of yellow fever with the slave traffic is clearly shown by a glance at our colored map; for nearly all the yellow spots are on our Atlantic seaboard and at northern points which were the first ports of debarkation for the slave ships during the period of the legitimate traffic covering more than a century because the centers of population at that time were all in the New England and Middle States. It is true the first authentic epidemics occurred in the West Indies in 1647 shortly after the arrival of slave ships there, and fifty years prior to its first appearance on our continent at Boston, in 1693, after England had begun to supply her American colonies

with African slaves. But it must be remembered
that Spain was the first civilized nation to use
Africans as slaves; and she first supplied her
West India possessions with them by contracts
with other powers, she having been interdicted
by Pope Alexander VI (1493) from going east of
a certain meridian. So as African slavery was
first established in the West Indies, this consti-
tuted them the *"fons et origo"* as they continue
to be till this day, the fountain head of yellow
fever; though as the early writers attributed the
first epidemics to the arrival of slave ships, they
thought its origin must be in the East. But
of all our yellowed spots showing where the
disease has prevailed at some time in the past,
there is not one spot of land among them shown
to be its birth-place or original home.

Now Dr. Audouard's discovery was that the
terrible epidemic at Barcelona in 1821, in which
five thousand persons died, was caused by the
arrival of two old slave ships, the Grand Turc
and the St. Joseph, with a cargo of West India
produce; and they were found to be so foul and
filthy that they had to be scuttled. At Passages
in 1823 he found an innocent looking brigantine,
the Dionostiarra, which had brought a cargo
from Havana and was then having her hull repaired. The first cases of yellow fever occurred
among the carpenters engaged in these repairs,
and they attributed their illness to a foul, sickening smell issuing from the hold and bilges of

the vessel. From these cases an epidemic followed; and inquiry into the vessel's history revealed that she was then engaged in the contraband slave trade and was completing her round
voyage, having taken negroes from Africa to the
West Indies and brought back a cargo of the
products of those islands. Dr. Audouard, finding that there were sixty or seventy ships engaged in this traffic making these round voyages,
argued that the same circumstances that he
found at Barcelona and Passages must have occurred repeatedly at the various other Spanish
ports where yellow fever had been epidemic time
and again. The historical fact that when these
vessels ceased to arrive at those ports yellow
fever also ceased to prevail and disappeared
from the coast of Spain to return no more, is
strongly corroborative negative evidence of the
truth of his argument.

The theory deduced by Dr. Audouard from
the occurrences at Barcelona and Passages was
that the specific infection which caused yellow
fever issued from the hold of the slave ships,
having been generated by the fermentation and
putrefaction of the dysenteric discharges and
other filth accumulating from the negroes
during the voyage over from Africa. So great
was this accumulation that the cleansing of a
slaver was said to be a Herculean task indeed
at which white men could not work, and the negroes were thus forced to do it. This was where

the traffic was carried on under government inspection and larger and better ventilated vessels were used. But the contraband ships were
small, ill-ventilated, over-crowded and never
properly cleaned or disinfected; so that it can
be readily perceived that they carried within
their walls enormous quantities of that peculiar
filth necessary for the development of the germ
principle of the disease, and that thus they became mediums for the transmission of the
materies morbi of yellow fever back and forth
over the Atlantic and were a menace and source
of danger wherever they cast anchor, even though
they came with clean bills of health and no cases
of the disease had occurred during the voyage.
Audouard and other writers upon this point
seemed to think it was necessary for this filth to
be thrown overboard and mixed with the mud
and filth of the seaboard cities in order for it to
produce yellow fever; but there is no reason
why under suitable meteorological conditions it
should not develop on board and become the
fever of the voyage. This is why I said it was
born upon the high seas; and taking into con-
sideration all the facts and conditions connected
with its production and comparing them with a
few well authenticated and similar cases, we are
fully justified in concluding that the first cases
of yellow fever with which our race was afflicted
originated on board slave ships during the voyage from Africa through the "Middle Passage;"

and this is the only rational explanation
enormous mortality among the white cr
these ships.

That this excessive mortality did occ
learned only by Clarkson's persistent i
and investigation of the matter; for the
edge of it was suppressed by the owners
vessels, as they feared that if it were ge
known it would prevent their obtaining
cient number of recruits to properly man a
the ship. This was an example of the su
sion of truth for mercenary and pecunia
sons. While Audouard's theory was supp
at first for political reasons, it was afte
suppressed or rejected by the physicians
United States, especially those of the
through bigotry caused by the passioi
predjudices arising out of the institut
slavery. That institution was maintain
guarded with so much jealousy, and our
were so deeply imbued with the theory
local origin of yellow fever as taught
great Doctors Rush and Physick, that we
patience to entertain or investigate any tl
which reflected upon or conflicted with t'
stitution. Born and reared upon a slave
tation, as were my father and grandfather
me; and familiar with all the workings of
stitution which was so long a curse and ii
upon our fair land, I feel privileged to
thus of the institution and its evil results.

To my mind there is no one single etiological
or pathological fact in the domain of medicine
that rests upon so sure and clear a basis for its
truth as the fact that this far-reaching Nemesis,
yellow fever, was the product solely of the nefarious slave traffic. This can be demonstrated
beyond the power of refutation, and clearly con-
stituted it an artificial disease resulting from
man's violation of natural law; hence it should
be and is in the power of man to exterminate it.
I claim that it has been driven to and is now occupying the last ditch, not by man's efforts, it
is true, but by the forces of nature working under favorable conditions. In this last ditch of
defense, or rather of existence, but little of man's
labor would be required to aid the same forces to
silently but effectually sweep the last vestige of
this scourge from the earth.

Look again at our yellow-spotted map of
the world and with unprejudiced mind trace
the traffic and mark the ports of most frequent
debarkation during the period when it was recognized as a legitimate trade and see its constant
and intimate association with yellow fever from
1693 down to 1808. Read the history of its frequent and destructive ravages in the most important of those northern ports, Boston, New
York and Philadelphia; then mark its sudden
and complete disappearance upon the abolition
of the traffic by England and the United States
in 1808. Observe how closely it accompanied

the contraband trade to the West Indies, where
its most destructive epidemics occurred during
the most prosperous period of the illegal traffic;
and how when this was suppressed by English
and American cruisers it was diverted to South
America; and how, as the slave trade of Brazil
enormously increased, yellow fever again made
its appearance in Rio Janerio for the first time
in a century and a half.

As stated before, the terrible epidemics that
ravaged the coast of Spain during the first quarter of this century were shown by Audouard to
have been caused by the specific infection issuing from the hold of one or more of those contraband ships which had arrived, shortly before
the epidemic began, with a cargo of West Indian
products on the return voyage from having carried a cargo of negroes from Africa to those
islands. Audouard in his investigation of the
epidemic at Passages in 1823 found that the
Donostiarra, a slaver, had arrived from Havana
with a clean bill of health and that there had
been no case of yellow fever aboard during the
voyage, neither had there been any in the port
prior to her arrival. Her cargo being disposed
of and her hull needing repairs, the carpenters
of Passage were employed to do the work; but
as soon as the first plank was ripped from her
bottom the workmen one after another began to
fall ill of yellow fever, and they rightly attributed their illness to a sickening smell that came

from the foul bilges of the vessel as they opened
them up. Will any unpredjudiced person demand clearer proof of the relation of cause and
effect as presented in the history of this case (and
it is not a solitary case either), or will any one
be so unreasonable as to require further evidence
of the relationship between the slave traffic and
yellow fever than I have adduced in the preceding pages wherein I have shown by incontro-
vertible historical facts that a most intimate,
constant, close and unbroken connection and
association existed between the two; both in
time and place, from their earliest history down
to the time when, the traffic being suppressed,
the cause ceased to exist.

Is it then unreasonable to hope, yea, to expect
and believe that in time the effect, too, may be
eradicated?　This has already transpired at
various times and places.　Just prior to the war
of American Independence the slave trade was
at its height.　England alone had one hundred
and ninety-two ships engaged in the traffic, with
space for transporting 47,146 negroes (Encyclopedia Britannica—"Slavery.") with which she
was supplying her North American colonies that
she was so soon to lose.　Thus during the legitimate traffic these vessels, with many of
other nations, entered our northern ports hundreds of times annually; and the peculiar filth
which gave rise to the specific infection of yellow fever was pumped up and thrown overboard

from their holds in material quantities t
and mingle with the mud of their harbors
penetrate the soil under the wharves and
lying quarters of the towns where, und
heat of a summer's sun, it fermented and
ated that specific poison, thus enablin
disease to become and remain indigenc
long as the yearly accretions of that pe
filth continued; and this afforded a r
basis for the opinions enunciated in ' 17
Doctors Rush, Physick and others of the
delphia Academy of Medicine that the d
was a native of our country. But see hov
after the suppression of this traffic the s
ceased to occur in those same ports where
century it had prevailed so frequently and
lently.

Water, one of nature's most useful
powerful forces, in the shape of the
lent, tossing, pitching waves and tid
the Atlantic Ocean was the means by
those polluted harbors were cleansed, the
ities being favorably situated for the accom
ment of this purifying process without
aid. All cause of the local origin of the d
thus having been removed the rigid enforc
of a perfected quarantine system has sinc
served them, with one or two exceptions,
any more of its ravages. This is only on
ample of the many furnishing incontestabl
dence of the exotic nature of the *materies*

of this specific disease and of the intrinsic value
of quarantine in preserving a country from its
invasions.

Yellow fever prevailed at Rio Janerio for a
number or years during the latter part of the
17th century, but disappeared, when the slave
trade was diverted to the North American Continent, and was unknown there for a century and
a half; but when the contraband ships, seeking
a market for their cargoes, after being driven
from the ports of nearly all other countries, began to arrive there in 1849, yellow fever also
accompanied them and established itself as a
disease new to the country, prevailing with unexampled virulence. It has remained there ever
since and is said to have become indigenous.

Personally I know nothing of the topography
of Rio or other Brazilian ports; yet it is fair to
presume that having disappeared once, naturally,
it may do so again or be made to do so by the
intelligent efforts of man. But the great
source of danger to our country is not the ports
of Brazil nor the Island of Cuba itself nor the
city of Havana in particular, but the putrid
water of the bay of Havana—an almost tideless,
stagnant pond dotted with mud-banks or shoals
and surrounded by high hills with but a single
narrow opening to the Gulf, so that its waters
cannot be changed. It has so long been offensive and full of decomposing organic matter that
as far back as 1821 there was a standing order

in the the British naval service not to use the water for any purpose. Doubtless the first poor negroes and slaves in the Western Hemisphere came in a Spanish galleon into that then unpolluted bay; for it is said that a few were brought over even during the time of Columbus, and that he recommended the exchange of Indian prisoners for Africans. Be that as it may, Havana was headquarters for the slave traffic from its beginning until its cessation; and for three centuries hundreds of slavers annually cast anchor in the bay, scoured and threw overboard the filth of their holds and pumped the putrid water of their bilges into it. So great was the quantity of this peculiar filth thus added to these quiet waters at times that it penetrated and became a part of the mud-banks, permeating even the soil of the shores, of the wharves and of the low-lying quarters of the town, vitiating and polluting the sea water to the extent of rendering it temporarily destructive to the fish.

I mentioned that Dr. Audouard found that the Donostiarra, to which he traced the origin of the epidemic of yellow fever at Passages in 1823, had arrived from Havana with a clean bill of health and had had no cases of the disease on board during the voyage. Innumerable instances of the same kind are on record of vessels having left Havana with clean bills of health, not a single case of yellow fever existing in the city at

the time of their sailing; yet they proved to be
sources of danger and infection to other places
which they visited months afterwards because
they took with them in their holds and bilges a
sufficient quantity of the necessary filth, obtained
from the water of the bay, to generate the specific infection.   A great deal has been said and
written about these Cuban "clean bills of
health," and Dr. S. E. Chaillé, President of the
United States Havana Yellow Fever Commission,
in 1879, writes in his report: "To the sanitarian it is a deplorable farce that commerce
should be burdened with such bills of health as
the Cuban authorities at Havana issued after
October 4, 1879."   But if Professor Chaillé and
the Cuban authorities had understood clearly
the true status of the case and known where the
real danger lay, he would not have had occasion
thus to characterize them for they would not
have been issued if the authorities had been
honest.   This, however, the professor questions
in no doubtful language; but let us leave the discussion of this point to future revelations upon
the subject.   ─

The great difficulty and hindrance to the ac-
quisition of a correct knowledge of this unique
disease, from the days of Rush down, has been
too much egotism and too great confidence in
the mind of each observer and investigator that
he alone was right and everyone else wrong in
regard to the respective theories and ideas ad-

vanced. This most singular of all peculiar diseases produces very strange but varying impressions upon the minds of persons in the different walks of life who have been so fortunate as to recover from an attack. Dr. LeMonnier in an article upon the subject written after the epidemic of 1873 says: "The more we physicians see of thedisease the more we desire to see of it and the greater we discover our ignorance to be;" and Chopin, of New Orleans, whose opportunities for investigation and observation during the epidemic of 1878 were unequalled, writes: "We really know nothing about yellow fever." But the most singular effect of all is that which it produces upon the minds of non-professionals. I do not think I exaggerate much in saying that nine hundred and ninety-nine in every thousand who pass through the disease successfully get up deeply imbued with the idea that they possess all knowledge of the disease; and in succeeding epidemics many of them present themselves as superior not only to all skilfully trained yellow-fever nurses but to the whole medical profession in their ability to care for and treat patients; and in addition some write learned (?) essays for the newspapers on the origin, nature, treatment, etc., of yellow fever. It is possible that the medical profession is responsible for this because of their proneness to over-medication, especially in a disease like yellow fever of the nature of which we are

totally ignorant and for which no antidote or
positive cure has been discovered. That Chopin
was practically right in 1878 and that we have
made no material advance in our knowledge of
this disease since the days of Rush, was clearly
demonstrated during the late epidemic of so-
called yellow fever which prevailed so exten-
sively throughout the entire Southwest during
the months of September, October and November, 1897.

If it were not the fact that but few physicians
of the present generation have ever seen a case
of yellow fever and that its visits to this country
fortunately are becoming fewer and farther between, it would seem passing strange that the
profession are still wholly unable to recognize
this unique disease which is so strongly individual in its characteristics and whose facies are so
remarkable that they should not be confounded
with those of any other malady. But I am
wandering from the real subject of this section
which I will close by endeavoring to recapitulate
the main points that I have presented.

I believe that I have established as a fact
beyond the possibility of refutation that there
has always been such a close and continuous
connection between the old African slave trade
and yellow fever as to clearly put them in the
relation of cause and effect. This being accepted as a truth places yellow fever in a cate-
gory by itself as a purely artificial disease, that

is, one caused solely by acts of man; hence it is possible to exterminate and eradicate it from the globe. The history of the cause, the slave traffic has been closed forever; while the continuous history of the effect, yellow fever, demonstrates the plausibility and great probability of the fulfillment of my prediction. It needs but a glance at our yellow-dotted map to show that, since the complete suppression of the traffic, yellow fever has entirely disappeared from the ports of Spain and all other places on the Eastern Hemisphere where it had prevailed during the existence of the slave ships. So also it has ceased to appear at the ports of our Atlantic seaboard; and it is only occasionally now that it slips into a few of our gulf ports owing to their contiguity to its last stronghold,- Havana, which has so long and so generally, though erroneously, been supposed to be its original home.

I have not been able to find an author who could ascribe its origin positively to any country. The United States and Mexico ascribe its origin to importation from Havana, while Havana and Brazil attribute it to Spain who in turn haughtily rejects it and refers the origin back to the West Indies, or to Asia and Africa; but these countries indignantly deny the charge and say that they never gave it to Europe, but always received it from her. So as no land has been found that will acknowledge the paternity of this nomadic Nemesis or accept it as a legitimate offspring, I claim my location of its origin to be the true one—born upon the high seas as a specific infection generated from peculiar filth found in the holds of African slave ships, these vessels being the medium by which this poison was transported to and distributed among the various ports to which their nefarious traffic directed them. - I find as valuable a testimony to sustain my position as I could ask, given, though unwittingly, by Prof. S. E. Chaillé in 1874. In his paper on "Vital Statistics of New Orleans from 1769 to 1874" he closes the paragraph on yellow fever, while yet arguing against the value of quarantine and in favor of the local origin of the disease, with the following: "The experience of the United States river fleet at New Orleans in 1863-64 confirms many other indications that yellow fever is especially prone to originate in the holds of vessels."

Now Assistant Surgeon Harvey E. Brown of the United States Army, in his report on "Quarantine on the Southern and Gulf Coasts," referring to this same "river fleet" of 1863-64, says: "The vessel on which the yellow fever originated, the Virginia I believe, had been an old slaver but was captured in the early part of the war and turned into a gunboat." What more direct and positive evidence of the truth of this proposition could be desired? While it

was a strange admission for one to make who
was even then teaching the local origin of the
disease, yet it is not a solitary case; for from the
days of Rush down to the present I find the
most ardent and zealous of the medical writers
who taught this theory qualifying their opinions
and reports by admitting that the arrival of
vessels from the West Indies or other known
foci of infection frequently caused a rapid increase of the fever, owing to the foul air and
offensive odors which issued from their filthy
holds and reeking bilges.

Surely they do not mean by these admissions
to claim that the holds of those nomadic buc-
caneers were a part and parcel of the port in
whose harbor they temporarily cast anchor; for
it was only in the holds of vessels of this class,
slavers, that the specific poison has ever been
known to be generated, with the notable excep-
tion of a few transports carrying negroes and
Chinese coolies and of certain wooden ships lying
up in West Indian ports. I will now give an
account of these, quoting from the writings of
Dr. C. Creighton to whom I am much indebted
for many of my foregoing thoughts and also for
recalling Dr. Audouard's theory to my attention,
which I had rejected and refused to properly
investigate forty odd years ago.

"The Regalia, a British transport, was employed in 1815 to carry black recruits from the
coast of Guinea to the West Indies. When on

the coast the health on the ship had been excellent, but during the voyage much sickness, chiefly of the dysenteric kind, occurred among the blacks. Thereupon yellow fever broke out with great malignancy, attacking all on board except the blacks who from first to last were exempt." The case of the Regalia is well known and it used to be quoted as showing that yellow fever was only a form of malarial fever, the malarial miasm in this case having come from a quantity of green wood that had been shipped at Boa Vista. The green-wood theory was always improbable, and the modern disentanglement of yellow fever from malarial remittents deprives it of whatever small probability it ever had. The other case which I take from Gillespie is equally suggestive. The French frigate, La Pique, fell into the hands of the English when Martinique was taken in 1794, and in November 1795 was sent with a prize crew to the Barbadoes. On the voyage they took two hundred negroes from a French vessel that was in danger of foundering. The negroes were confined in the hold and in a short time yellow fever appeared among the La Pique's crew and proved fatal to one hundred and fifty of them, although it did not attack the negroes at all. "Such a mixture of men, strangers to each other," says Gillespie, "has often been found to occasion sickness in ships; and together with other causes operated fatally before the

arrival of the ship at the Barbadoes. * *
is a melancholy instance of the generatic
fatal epidemic on ship board at a time w
inhabitants of the Barbadoes and the crew
other ships in company with the La Pi
mained free from any such disease."

It will readily be seen that Audouard's
of the immunity of the pure-blooded negr
yellow fever is perfectly sustained in boi
instances. As still further testimony.
truth, Dr. Creighton, referring to the
expedition to Mexico, and says; "The
regiment of five hundred blacks, recrui
the French in the Soudan and Nubia
expedition, did not lose a single man no
have a single case of yellow fever in t
demic at Vera Cruz in 1866, while the
soldiers were dying by the hundreds."
cites the circumstance of the product
yellow fever on the coast of Peru, in 1
the importation of Chinese coolies, an
"Although apparently in contradiction
facts are really corroborative, for the c
themselves were as exempt from the
there and then as were the negroes on th
site side of the continent, having been t
there under the same conditions."

Of wooden ships Dr. Creighton sa'
shall mention briefly the most terrible hi
instance of ships charging themselves w
poisonous mud of a slave port. On the c

of Port au Prince, Hayti, June 4th, 1794, about forty merchantmen were found in the harbor, most of them large vessels laden with cargoes of coffee, sugar, cotton and indigo which had been lying stored in them from one to three years. Owing to the suspension of business during the Revolution many of them had never had their holds open all that time. English prize-crews were put on board to navigate them to Port Royal and other British West Indian ports, and they had hardly put to sea when yellow fever attacked them with unheard-of sud-denness and virulence. One of the prizes was picked up by a Guineaman and everyone on board was found to be dead. Even the negroes who were put to cleaning them out took the fever and died."

I will now go back to the fierce controversial-ists in the last decade of the eighteenth century at Philadelphia, for I promised to explain my paradoxical remark that both parties were right in those angry controversies. I verily believe that if those intellectual giants in medicine, Doc-tors Rush, Hutchison, Physic and others of the Academy of Medicine on the one side, with Doc-tors Curry, Carpenter, Chapman and others of the College of Physicians on the other, could revisit the scenes of their labors and see the present status of the two, which I call cause and effect, i. e., the slave traffic and yellow fever, they would all be enabled easily to see not only

wherein they were right, but also where
were wrong and would grandly confes
errors, unitedly agreeing that my ded
are rational and that they are sustained
the facts in the subsequent history of the (

First. I have shown that from the tim(
arrival of the Dutch ship with a few Afri(
the Virginia coast in 1620, down to 177(
sibly to 1793, the peculiar filth to which (
ease owes its origin had been gradually de
in various harbors of our Atlantic coast, e:
ly in the mud of the Delaware at Philad
till a sufficient quantity had accumulated
der the disease temporarily indigenous, s
ing Doctors Rush and his party in claimi1
it did originate there after 1793. But, as
said before, I find that they nearly alway.
fied their reports by referring to fresh :
of infected vessels as adding greatly to t
of the disease.

Second. The importationists, Cur1
his followers, were right. This party
ways right everywhere, for if my loca
the birthplace of the scourge is correct,
originated *de novo* on the high seas, it nec(
follows that it was originally an exotic
countries and had to be imported.
Philadelphia this party erred in denying
ever originated at all, although they could
be blamed, for it was never, *defacto*, a
there, as the material for its generati(

brought in the bottoms of transient rovers of the
seas and deposited there.

Third. They would readily see that the disease has entirely disappeared since the suppression of the slave traffic, and that the waters of
the Atlantic, unaided, had long since removed all
traces of its peculiar filth from every port on
that coast where it had so frequently prevailed
in their day, from Portsmouth, N. H., on the
north to the cape of Florida on the south, and
that a rigid quarantine had barred its entrance.
Yellow fever has since also disappeared from
our gulf ports as well, New Orleans probably being the last point to yield because of
later additions to the filth at that point; but
the immense volume of water annually poured
down the mighty Father of Waters has finally
been victorious and swept the last vestige of it
into the sea, so that it cannot be truthfully
claimed to be indigenous to any part of the
United States.

Fourth. They would also see that it has
ceased in all places in Europe and Africa where
it prevailed almost constantly during the existence of the slave traffic, as well as in all the
ports of that hot-bed, the West Indies, except-
ing the foul bay of Havana; and they would not
say, after a glance over our yellow-dotted map,
that I am using a fanciful or extravagant expression in saying that it now occupies the last
ditch, in a forlorn hope of continued existence

YELLOW FEVER.

from whence it occasionally does send c
tachments of germs of the specific infect
fomites of various kinds in filthy vess
ravage certain localities, but that it cann
never will again pollute another harbor
world sufficiently to enable it to beco
indigene in that country.

That "ditch" can easily be cleaned by
ligent efforts directed to aiding Nature's
minister of cleanliness, water, to proper
form its task of ablution, by digging 1
more canals from different points on th
(even across the island if necessary, for e
should not be counted in such an ente
opening into that sluggish bay, in or
create currents sufficient to completely
the water and carry out all the fermentii
putrifying organic matter contained t
Then should follow proper cleaning and
tific disinfection of the mud of the shor
low-lying parts of the town, until
trace of the poison be eradicated; and
same time all negroes should be remove
these localities.

I know this will never be done by the
owners of the Island; but it can and sho
done for the good of humanity. Two pe
and highly civilized nations like Englai
the United States, by assuming a prote
over Cuba for Spain, (I doubt if the
population is worthy or capable of self-

ment) could easily accomplish this and at the same time put a stop to the inhuman butchery now going on (the winter of '97-98), which is a disgrace to the nineteenth century.

Let all harbor revenues above the expenses of the protectorate be applied to this one object and it would soon be accomplished. Then the work of this Nemesis will end, its mission of vengeance having been accomplished, and yellow fever will become as truly a thing of the past as has its cause, the slave traffic; while at the same time the value of Cuba will have been enhanced an hundred-fold.

One thought more. I have stated the error of the importationists; that of the other party was two-fold, both in denying the possibility of the transportation and importation of the *materies morbi* of yellow fever, and still worse in attributing its origin to local filth, the putrid exhalations of alleys, gutters, docks and stagnant water near the city, thus making it belong to the class of paludal or marsh fevers—an idea overflowing with error and danger, yet still adhered to by many. In Memphis, 1878, thirty out of forty-five young, healthy, stout, volunteer physicians sacrificed their lives to this hideous error. The transportability and importation of yellow fever has long since been definitely proven; and it is as distinct and different from malarial and other epidemic fevers of this country as smallpox is from measles. While I admit that it is pre-eminently a disease of filth, yet I agree fully with my lamented and revered teacher the late Professor H. F. Campbell of Augusta, Ga., when he said: "We may accumulate filth, piling it up from the pavement to the sills of the second-story windows, and it will never enable us to manufacture a single case of genuine yellow fever without the specific germ, any more than we can manufacture a case of smallpox without the specific virus of that disease."

Dr. Campbell had a clearer idea of, and understood yellow fever better than any modern investigator of the disease; and only lacked the certain knowledge that the disease owed its specificness to a specific infection generated from the peculiar filth of another race and found in the holds of old African slave-ships to make him perfect master of the whole subject.

PART II.

NATURE, ANATOMICAL CHARACTER, SYMPTOMS, COURSE AND TREATMENT.

I shall have but little to say under the above headings; for while I claim that we have more exact and positive knowledge, and a clearer comprehension of the etiology of yellow fever than of any other known disease, yet I admit that we are still groping in the dark as to its nature and cure. We know that it is a specific fever of one paroxysm lasting from thirty-six to seventy-two hours, and that it can kill or cease in these few hours. I maintain that the cause of this fever is a specific infection engendered from a fermenting, putrifying mass of the natural excreta and morbid discharges of another race occupying a lower order in the scale of the human family. Of the physical and chemical nature of this poison nothing is definitely known; but reasoning from its observed phenomena, which are constant, we may safely conclude that it is an organized, disease-producing germ—a particle of living matter from the living matter of the organism of that lower race, transportable in fomites of various kinds, capable of growth and multiplication to an indefinite extent, and, in suitable conditions, generating the true pathogenic microbe of yellow fever, its last stage of existence. Hence these microbes may hibernate and survive the winter, under favorable surroundings, and be revivified by returning warm weather; but they have no power of reproduction and can give rise to sporadic cases only, never to an epidemic. When this specific germ is introduced into a locality where the conditions are suitable, it takes up its march along the surface of the earth and begins its function—rapid reproduction and generation of the fever-producing microbes which rise and fill the air of the locality. From repeated observations made by myself and many others in numerous epidemics, it has been found that the germs travel about forty feet a day and that nothing impedes or stops their progress, being apparently governed and controlled only by the earth's lines of magnetic force in circles.

I have never entered an infected district but it recalled an incident of my boyhood days in which I was teasing an immense rattle snake with a long fishing pole, amused at the reptile's ever-increasing anger. It finally became so enraged that, after inflicting its own death-wound, it exhaled from its fangs a powerful, sickening odor which quickly filled the air and in a few seconds so affected me that I seemed to be standing in a circle of suffocative yellow haze. Dropping my pole, I ran for pure air till exhausted I fell almost fainting upon the ground and vomited freely. The odor perceived in localities infected by yellow fever and described by many as being of a fishy character, never so impressed my olfactories, but was the rather a constant reminder of the dissecting room of my college days. A person soon becomes so accustomed to this odor as not to perceive or notice it after being in the infected region a few days; just as one who smokes does not perceive the odor of tobacco in his room, while it will be very offensive to one entering who does not use tobacco. This odor of yellow fever, as well as the microbes, seems to be governed by the same mysterious force as that which controls the march of the germs; and there is often a well defined line of demarcation between the infected and the pure air, readily recognized by the odor above mentioned.

During the epidemics of 1878-79 in Memphis, Tennessee, while gazing upon the icteric and bronzed countenances of the victims of "yellow-jack," I was continually reminded of the mingling of the blood of the white and black races, possibly because I was then earnestly engaged in studying and investigating the long neglected theory of Dr. Audouard; and my mind may have been so influenced by his ideas as to cause me to view everything through the medium of his

theory, just as yellow-fever experts ar
"times charged with seeing all kinds o
through a yellow haze and diagnosing
be yellow fever. The fear that my min
have been biased in this investigation has
me to delay this writing from year to
twenty years; and now it is almost too late
count of my feeble health and failing]
and mental vigor, for me to expect to se
this idea the recognition which- it d
It is a task for me to write now, for mer
failed to such an extent that I have to
entirely upon records, and they are in su
fusion that I fear I shall fail in making i
ing and useful reading of them. But t
ject has lost nothing by this delay, while
gained experionce and, I trust, a moiety
dom along with the additional historic
and events of these twenty years that
and strengthen my position and my pre
made at that time.

    To return to the subject proper: Yellc
is not in its nature contagious in any ʂ
the word; and while, we say- it is an in
and epidemic disease, yet I must insist t
so only in a limited sense, and not like ɩ
la grippe, dengue, the plague, etc., v
times sweep, with greater speed than th
over extensive sections. It is well knoɩ
yellow fever, with rare exceptions, never
from an infected town or city into the su

ing country, but is frequently confined to a tier
of blocks or limited to one side of a street. The
only rational explanation of this peculiarity of
the disease is that which I have already given,
that it must be under the control of the earth's
lines of magnetic force of which we as yet have
a very vague and indefinite knowledge, despite
the immense advances being made in the control and use of electricity and magnetism during
the closing decades of this century.

The profession, at least those who have seen
and studied the disease, seem to be a unit upon
the question of contagion; but it strikes me that
their ideas and the ideas of sanitarians generally, more especially those in charge of our quarantine affairs, must be vague and unsatisfactory.
If it is truly a non-contagious disease, I ask why
that unknown and unsolved question, the period
of incubation, should cut any figure in the
matter of quarantine, there being no possibility
of contagion from the person. Now I want it
understood that I am an earnest and zealous advocate of rigid, individual state quaratine based
upon true, scientific principles; and in the case
of yellow fever, upon its two, prominent and
generally admitted characteristics, viz., the undoubtedly easy transmissibility and receptibil-
ity of the atmospheric germ, and the entire absence of contagion in the subject.

A quarantine based upon and administered in
accordance with these principles will forever

prevent a recurrence of that relic of barbarism,
"shot gun quarantine," the inconvenience and
tyranny of which I experienced several times
last fall (1897) when it was established all over
this broad State; and this all on account of the
insane and frenzied panic resulting from a few
cases out of many thousands of that painful but
innocent disease, dengue, having been diagnosed
as yellow fever by certain officials who ought to
have known better. The recovery of every case.
so diagnosed in Galveston and Houston may be
considered *prima facie* evidence, in the absence
of all other information, that the diagnosis was
incorrect. For we have the unanimous testimony of all past observers and writers upon
yellow fever that nearly all of the first cases in
every epidemic of true yellow fever die. There
are good reasons why this is so: first, because
the peculiar poison always attacks those of fee-
blest constitution first; second, because of the
inability to make a correct diagnosis till several
deaths have established the identity of the diseases.

It is folly to claim that the small mortality
of the so-called yellow fever in the South in
1897 was due to the great improvement in modern treatment and a better understanding of the
disease; for it is an undeniable fact that not a single case diagnosed as yellow fever in Texas during that time had the slightest advantage of
that wonderfully improved treatment of the *fin*

de siecle. And I do not hesitate to say that this
much vaunted improvement in treatment is a
myth, and that if we ever have another serious
epidemic of genuine yellow fever in this country, which I think is extremely doubtful in spite
of the many dire prophecies for this year, there
will be the same old death-rate of from twenty-
five to fifty per cent.

But to continue the subject of quarantine:
The so-called period of incubation has been
reduced gradually by these sanitary scientists during the past twenty years from forty days to five.
And if it could be demonstrated beyond doubt
to even be one day or one hour only, again I
ask what it has to do in the question of quarantine. As I have clearly shown, centuries of
observation and thousands of experiences have
fully demonstrated the perfect harmlessness of
the unfortunate refugee who seeks shelter and
hospitality at our hands in his hurried flight
from this deadly pestilence. But this statement
will require to be more fully understood before
it will be accepted even by the modern sanitary
scientists, who profess to believe and do teach
that yellow fever is not contagious, but who yet
enforce a quarantine and maintain camps for the
detention of unaffected persons, which act flatly
contradicts all their teachings and affords
rational grounds for the people in the country
to establish "shotgun quarantine." These observations are of times prior to this fast period

of countless railroads and rapid transits, when
the common people apparently understood
yellow fever better than the medical profession
of today, experts included; for they knew it was
not contagious and did not hesitate to receive
the fleeing refugees with open arms, caring
tenderly for the sick and closing the eyes of the
dying, confident that there was no danger to
themselves. And they knew, too, that when
the specific exotic poison was introduced into a
city where it could live and propagate it could
not be confined to a house, or block by all the
powers and ingenuity of man.

Why in those days yellow fever was so
rarely, if ever, spread by the refugee was due
mainly to the slow methods of traveling by stage
and private conveyance, so that by the time he
reached his destination the germs contained in
his wearing apparal and baggage were all dissi-
pated or destroyed by nature's other great purifier, the air; and though his blood was teeming
with the poison sufficient to finally produce the
disease, with black vomit and death, it has been
fully demonstrated that he could no more have
communicated the disease to others by any
emanations or excretions from his body than
one poisoned with strychnine could communi-
cate his terrible tetanic convulsions to his
attendants. As it is clearly apparent that it
would require a sufficient amount of strychnine
to produce the same poisonous symptoms in an-

other, so it would require a fresh supply of outside atmospheric disease-germs to produce a
case of yellow fever among those surrounding
the bed of the dying refugee.

But in these days of rapid transit the refugee
may pack his trunk in an infected city, take a
fast train and travel one, two or three hundred
miles in a few hours; and while there is not a
particle of danger from his person, although it
may be filled with the deadly poison, yet not
only his apparel and his trunk but the uphol-
stered car in which he comes may be full of the
fever-producing germs and become foci for the
spread of the disease, if suitable conditions prevail in that locality. This explains the unprecedented extension and spread of yellow fever,
during the past twenty-five years, into the
interior cities and towns from the seaboard
cities that have always been considered its
natural habitat.

All the foregoing statements have been proven
to be true, times without number; if it were not
so it would be the extreme of stupidity and folly
to say that yellow fever is not contagious. But
notwithstanding these facts, the system of quarantine established and enforced in this country
is a complete stultification of our professed
belief and teachings.

With a clear understanding of these well-
established facts of the nature of yellow-fever
poison, is it not possible for scientific sanitarians

to formulate and put into operation, at far less
expense than the present incongruous and
absurd system incurs, a modified quarantine
based upon rational principles, and one that will
command respect and obedience because it will
no longer contradict that important and generally acknowledged principle, the impossibility
of personal contagion? This system would do
away with the expensive camps of detention and
observation; stop the trains from an infected
city five miles from the quarantining city; require the refugee to take a bath, clothe him in a
fresh suit supplied for the occasion, and allow
him to go to his destination (even though he be
then suffering with the fever) in other cars sent
for the purpose. The train bringing him would
then be returned to the infected city and his
own clothes and trunk with its contents would
be thoroughly ventilated and sent after him
as soon as possible. I say ventilated instead
of disinfected, the term now in vogue, because in this era of new things and new
methods, the sanitarian, in his search for
and experiments with chemical germicides or
some new artificial disinfectant, seems to have
forgotten the only reliable germicide of olden
times, nature's great disinfectant and purifier,
the air in motion.

We have seen that in consequence of the
slow modes of open-air travel in the past the
clothing and other fomites of the refugee were

purified, all the germs in them being either dis-
sipated or destroyed; thus were afforded thousands of instances of positive proof that yellow
fever was not contagious, and this was recognized by the people at large as well as by the
profession. With such a system of quarantine
for the interior, wisely and honestly administered, confidence would reign once more among
the people; and the frightful frenzies that we
have lately witnessed would give place to the
common sense and common humanity exercised
fifty to a hundred years ago.

What I have said has been in reference to in-
land quarantine only, and for the purpose of
showing the inconsistency between our professed belief and our actions, and also to con-
trast the present conduct of the general public
with that of fifty years ago. I have no experience in coast quarantine whatever; but if my
ideas and theory of yellow fever are correct (a few
more years will either verify or prove them false),
the point, for quarantine inspection, at which to
prevent the importation and introduction of the
infectious germs of this now clearly proven exotic disease would properly be at its natural
habitat. But this can be accomplished only
through an international compact between
Spain, the United States and other interested
nations.

It seems passing strange that such distinguished and able investigators and writers on

yellow fever as Doctors LaRoche, Chaillé, Fagét, Ford and many others should come so near
the truth and yet fail to grasp the key to the
many perplexing mysteries and semeiological
phenomena of this disease; for what Creighton
said of LaRoche is applicable to all of them:
"I find nowhere in his pages any evidence that
he had mastered the facts of Audouard's argument or duly weighed its conclusions, thus failing to catch the sparkle of the gem for which he
was seeking". These writers frequently mentioned the proneness of yellow fever to originate
in the holds of vessels; but this was too general,
and they thus failed to catch Audouard's idea
that it was in *special vessels only* in whose holds
was found that peculiar filth of another race
which breeds it. Had Professor Chaillé clearly
understood this I do not think he would have
labored so ingeniously as he did to prove that
quarantine was ineffectual against what he then
thought was an indigenous disease, because the
rigid military quarantine of 1864 did not prevent
the occurrence of yellow fever in New Orleans
that year.   .

But Professor Fagét testifies that the quarantine was against vessels in the merchant
and not those in the naval service; while
Dr. Fehner says: "Vessels of war were rigidly
quarantined, but those for transportation were
not." The decision of this point is rendered
immaterial by the light which later develop-

ments throw upon the true origin of the yellow fever of that year; for it was found that there had been a vessel lying at anchor in the river since spring, which had within its depths that peculiar filth that under suitable conditions engenders the specific infection of the disease. I refer to the gun boat Virginia before mentioned, an ex-slaver which, Dr. Brown says, "had been captured in 1863, shortly after landing a cargo of Africans on the east coast of Cuba, taken to New York, turned into a gun boat, and her hold thoroughly cleansed, when it was found that her hull was iron, with double walls and that the interspace being inaccessible was never cleansed. As a consequence the bilge water became so foul from the impurities contained therein that the pumps could not be used. However she remained a healthy vessel till the fall of 1864, when in September she was hauled onto the river bank for repairs, pumped out dry and cleaned." As a natural sequence to this proceeding came an outbreak of yellow fever.

The many acrimonious controversies that have occurred about this disease were the result of the great difference of opinion as to its true origin, the nature of its poison, its portability, contagiousness and period of incubation, and also, as it would now seem, the lack of knowledge of the whole truth about the disputed points, on the part of both parties to the debate.

Prof. Chaillé, after summarizing, under three heads, the various theories of yellow fever held by the profession, adds a fourth, a part of which is as follows: "Yellow fever is produced by two poisons, or, rather, by one which produces different results depending on the place where it is generated, the quantity generated, and the different conditions of these poisons. Thus the poison sometimes is either contagious or portable, or both, and at other times is not."

Now, while yellow fever is said to be identical in the various climates where it prevails, it is very natural to suppose that the nearer it prevails to its origin the more virulent and destructive will be its effects. It is in this way I account for its unexampled fatality and enormous rate of mortality in Spain; for her contraband ships could not be repaired or cleaned in any other ports; and after discharging their cargoes of West Indian and Brazilian products, which, have themselves been proven to be splendid fomites for carrying the germs of the disease, they sometimes lay from three to five weeks in the harbor, undergoing repairs, scouring their foul holds and pumping their filthy bilges into the water of the bay.

The rate of mortality in yellow fever has been less in the United States than in any other country, because the poison was more diluted and weakened in our harbors; and, as I have shown, since the complete suppression of the slave trade

they have been thoroughly cleansed of every trace
of it by the action of the waves and tides of the
Atlantic, so that yellow fever has ceased to re-
cur in them as an indigenous disease.   Hence
my prediction that it is nearing the end of its
existence and can be eradicated by the properly
directed efforts of man.   If this hope is fulfilled,
it will prove the correctness of the views of those
who now hold that yellow fever is a specific disease *sui generis*; but I find Prof. Fagét, in 1870,
exclaiming against this expression found in
Prof. Atken's "Practice and Science of Medicine," saying: "Yet, it has no characteristic or pathognomonic sign; black vomit even is
not such for it.   It is only by the agglomeration
and co-ordination of its symptoms, by its origin
and modes of importation, development and cessation, by its march especially, and by some other
peculiarities, that it may be distinguished from
other types; it thus essentially constitutes a
truly morbid species."   Strange words, truly, to
come from him who taught the profession that
a regular decrease of pulse with an increasing
temperature from the first or second day to the
fourth was an unfailing pathognomonic symptom of yellow fever not found in any other febrile disease, and who has further taught that it
strongly individual in its characteristics.

But as surely and certainly as the historical
facts and truths of the disease revealed in its
nineteenth-century record demonstrated that

yellow fever is not a morbid species, but a disease *sui generis* and peculiar, so surely is time proving that this much-relied-upon pathognomonic sign is no more peculiar and characteristic of it than even are black vomit and jaundice. While for twenty-five years I have relied largely upon this sign, but more upon the fact of its being a fever of one simple paroxysm, as an aid to diagnosis, yet I have always been convinced that there is no one single pathognomonic symptom by which the observer may be enabled to say positively, "This is yellow fever." The entire group of disease phenomena alone furnishes the key to a correct diagnosis, and which when analyzed exhibits the strong individual characteristics of yellow fever.

The period of incubation is a myth, and being an unknown quantity is of no value in diagnosis. There is no need of an incubating nidus in the organism; for the germs incubate in the atmosphere, and every person in the infected area inhales and swallows myriads of these disease-producing bacilli. But Pasteur proved that these micro-organisms cannot infect the body while the vital equilibrium is perfectly undisturbed (a condition unfortunately very infrequent); hence the great difference in time in which different individuals in the same epidemic succumb to the poison, owing to the individual powers of resistance; some escape its effects altogether.

I myself passed unscathed through half a

dozen epidemics, and when I finally succumbed it was because my vitality was below its normal. But my attack was an unusually light one, fever lasting only thirty-six hours, with but little cephalalgia or other pain. I left my bed on the fifth day and was out seeing patients on the eighth. This imprudence, with over-eating and the care of five children suffering with the same disease, caused a relapse eight days later which put me in bed for six weeks. On the day of relapse I returned home from making calls at three o'clock p. m., feeling well but a little jaded; ate a hearty dinner and lay down for a short nap. I must have slept heavily; and as I had been losing sleep my wife did not allow me to be disturbed till 5:30 p. m., when she came in with a lamp and aroused me. In that heavy sleep of two hours and a half I had become deeply jaundiced; hemorrhage from the nasal and buccal membranes had commenced, and I had swallowed a considerable quantity of blood which so nauseated me that on awakening I vomited freely and threw up such a large quantity of dark colored undigested food that my wife and my friend, Dr. A. H. Ketchum then fresh from college, thought I had the black vomit and so reported; but it was not true, and I never vomited or was nauseated again during my six weeks' confinement. There was never the least sign of fever and my pulse continued to descend until it reached forty per minute, then, after I was able

to be out again, slowly ascended to one hundred and oscillated back and forth for months
before it became steady at seventy-four, ten beats
less than my normal pulse was before the attack.

I had a narrow escape and only speak of my
case at this length because so many lose their
lives by the same imprudent course that I pursued; and I earnestly desire to impress upon
physicians that it is our duty to warn ever yellow-fever patient, no matter how light the fever
has been, of the great danger and risk to life
from getting up too soon and indulging the appetite; for, while many escape without further
trouble, I have known patients, who were being
congratulated on their quick recovery, to die as
early as forty-eight hours and as late as six weeks
after the fever. The one, two or three days'
fever, during the lifetime of the microbes as they
ravage the life-stream and afterwards pollute it
with their carcasses, is by no means the whole
of the disease. It is really only the first stage.
The calm following the febrile stage is a very deceptive one and is often a more or less profound
state of collapse, the result of the destructive
changes wrought in the blood-stream by the poison, and often escapes the eye of the busy, over-
worked practitioner until it is too late. It is
during and after this stage of calm that organic
changes begin; and their extent and gravity,
with the consequent depression of vital processes, depend upon the extent of blood-

changes and the amount of paralysis sustained
by the great sympathetic.

But to finish the subject of incubation: Dr.
Dowell and other authors on yellow fever report numerous instances of persons who, having
fled from an infected city to the country, or a
distant town, were attacked from three to six
weeks afterward; and these authors adduce these
facts as positive proof that the incubation period
may be that long. But this can be more ration-
ally explained by the now well established fact
of the easy portability of the disease-producing
germs and microbes in the wearing apparel and
baggage of the refugees. These people packed
their trunks, generally of the largest Saratoga
pattern and several of them, in the infected city,
thereby enclosing a portion of the disease-laden
air with their winter garments, which they always carry knowing they will not be able to return till after a freeze. The late development of
the fever in them was not due to incubating
germs inhaled while in the city, but to their in-
haling those enclosed microbes, *"compagnons
de voyage,"* upon opening their trunks at the
periods mentioned. Their nervo-vital powers
having been lowered by the excitement, fright
and fatigue which they had undergone, they fell
an easy prey to the action of the poison. Numerous instances of this kind are on record
which have been thoroughly investigated; and
abundant evidence, too plain and convincing to

be gainsaid, has been found to prove that my explanation is the correct one.

Reasoning from analogy, I am of the opinion that when this deadly animal poison gains an entrance into the blood-stream, it begins its work at once and acts as rapidly as the venom of the rattlesnake. In every epidemic I have seen numerous instances in which persons were stricken down within twenty-four hours after entering the infected atmosphere; and the reason that every non-immune is not thus affected is that the vital energy of the lungs and alimentary canal is sufficient to resist the entrance of the microbes into the circulation and also to destroy them. As soon however as a breach is found in a weakened organism the microbes rush in, rapidly fill the blood-stream, and the work of destruction and death begins. Instantly all is confusion and terror; and urgent, pleading messages for help flash over the nerves from every organ and tissue to the great central organ of nervo-vital energy, the brain. So numerous and rapid are these messages, saying that function is impossible with such a blood supply, that the poor brain, suffering from the same cause also, becomes so irritated, confused and helpless that in dispair it flashes to the outer world the concentrated impressions of the entire organism, and there appears upon the countenance an expression of surprised horror, followed quickly by one of fierce inquiry in which the eyes seem to pierce the inmost soul of the physician, seeking help and hope; and seeing only dispair there, the expression of that emotion spreads over the countenance as the last ray of expiring intelligence, and the whole frame becomes convulsed. The dumbfounded physician and attendants can only watch, powerless to help. Convulsion succeeds convulsion till at last the material body of the miserable victim exhausted dies, exhibiting all the phenomena of putrid decomposition.

The expressions of the emotions of surprise and horror appear so nearly simultaneously that I designate the combination as "surprised horror." I have never seen this expression of the countenance in any other disease; still it cannot be strictly considered pathognomonic, except in a limited sense, as it is seen only in those rapidly fatal cases constituting about ten per cent of the whole number. I mention it because it is only one of the *ensemble* of expressions that go to make up the peculiar facies of yellow fever. I have personal knowledge of its appearance being deferred in quite a number of cases till that deceptive stage of calm just after the subsidence of the febrile stage; and in every instance the patient was sleeping quietly, and the expression was produced by the shock from being suddenly awakened by a loud noise, as the blowing of a steam whistle or the ringing of a church bell near by.

One such case was that of a man who occupied a room in the second story of a building on
The Strand, Galveston, Texas, during the epidemic of 1859. He had passed through the fever
safely and, though recognized as very ill, hopes
were entertained of his recovery. He slept quietly through the night till 3:00 o'clock a. m.,
when a Trinity River steamer arrived at the
wharf just under his window and began to blow
its discordant whistle. At the first blast the patient sprang up on the side of the bed and with
hands over his ears said, "If that whistle does
not stop in five minutes, I am a dead man."
Knowing the custom of the boat to blow half an
hour, a messenger was hurriedly sent to have the
noise stopped; but before it ceased that horrible
expression appeared upon the sick man's countenance and was followed by a convulsion which
soon closed the scene in death.

Another was that of Mrs. H., Calvert, Texas,
1873. She passed through three days' intense
fever but did not exhibit the amount of prostra-
tion usually seen in the stage of calm. I saw
her at midnight, twelve hours after the fever
had subsided. She was feeling so well that she
insisted that her lady friend who was nursing
her should go to bed and let her thirteen-year-
old daughter lie by her on the bed to call if anything was needed. She slept quietly till 5:00
a. m., when a switch engine, standing upon the
track forty or fifty feet from her window, having

gotten up steam, blew a sudden shrill whistle
which instantly awakened her and she sprang
out of bed in great alarm, with eyes wide open
and wildly rolling as if she saw all the hob-gob-
lins ever imagined. Her little daughter pushed
her back upon the bed and called for help. Living only a block distant, I was at her bed-side in
less than ten minutes; but alas! those invariably
fatal convulsions had begun, and they continued
until she died at ten o'clock that morning.

That common and annoying nuisance, the loud
and long blowing of whistles, should be strictly
prohibited by law during an epidemic of yellow
fever; for it is well known that many a poor suf-
ferer, with life trembling in the balance, has had
the scale turned against him by their unearthly
and at all times needless screeches. In this
city of railroads, Houston, Texas, where some
fifteen or twenty either center or pass through,
with its several oil mills, ice, beer and other fac-
tories in active operation, discordant and disa-
greeable steam whistles are to be heard almost
constantly day and night, often preventing con-
versation for the time being. This should not
be permitted at any time; for there are nervous
sick people all the time, and the shocks from
these piercing, ear-splitting sounds is very inju-
rious to them.

I said these unfortunate patients die with all
the signs of putrid decomposition; and this is
the only anatomo-pathologic feature or charac-

teristic of any importance that I shall notice in
detail; for in my search for diagnostic signs I
have been in the habit of studying the disease
in the living subject in preference to the cadaver, and consider it a waste of time to study
the useless natural history of the latter. The
above pathological fact was first brought to my
attention years ago by an observing undertaker,
in the case of one of my patients who had died
within forty-eight hours after the attack began.
He said to me, "Dr. I want to get that man into
his coffin just as soon as possible, and if you
will go with me I will show you how very rapidly decomposition goes on in those who die so
quickly of yellow fever; you can explain to the
family my apparently indecent haste." I found
that he had to use great care in washing the
corpse lest the skin should slip off; and the
muscles had a peculiarly soft, doughy feeling
which produced the impression that with a little
more pressure the fingers would easily penetrate
the yielding mass. The odor of putridity was
exhaling from all parts of the body. Since then
I have closely observed all such cases, and have
invariably found decomposition of a putrid character progressing far more rapidly than I have
ever seen it in any other disease, both before
and after death. This decomposition, the result
of the profound blood changes wrought by the
poison, undoubtedly begins with the first entrance of the microbes and is continuous until

the fatal termination; then of course it becomes
more rapid, all vital resistance having ceased.
This rapid decomposition is greater and more
readily observed in all those quickly fatal cases
in which the system is, as it were, overwhelmed
by the immense quantity of poison entering it;
still it occurs in all cases, to a greater or less
extent, even in those who recover.

This tendency to rapid decomposition is denied by some observers, and admitted by others.
These latter, however, go a step too far in deducing the theory that this rapid decomposition
constitutes a condition which causes a rapid increase of bacteria, fungi and other simple organ-
isms that become carriers of the poisons and
thus aid in disseminating the disease. This is an
utter impossibility, for there is no longer any
yellow-fever poison in that dying or dead body.
The specific, yellow-fever microbes perish in the
system amid the wreck and ruin wrought by
them just as quickly as does the rattlesnake
from its own venom self-inflicted. If this were
not true we should be compelled to admit that
yellow fever is contagious from the person.

I have repeated this because I want the profession to get a clear understanding of the idea and
to become familiar with it; for it was a long time
before Prof. H. F. Campbell could get this
truth to penetrate my dull brain so that I was
able to thoroughly understand it; and without
the proper comprehension of the whole idea it

is impossible to comprehend the non-contagiousness of yellow fever from the person, when there
are so many instances on record of an epidemic
having started in an interior town soon after the
arrival of refugees from an infected city. Even
Prof. Chaillé admitted that he found it difficult
to comprehend the distinction apparently made
between the non-contagiousness from a patient
in his clothes and the portability of the disease
by the same clothes. But I would be considered
prolix and tiresome if I again repeated what has
already been mentioned twice in the preceding
pages, and I hope my readers are now able to
fully and clearly understand the point in question.

It would seem that the anatomo-pathologic
lesion above mentioned ought to be sufficient
under this head; but I will mention briefly two
others which I observed frequently while making some investigations in Memphis in 1879. Of
twenty-five cadavers the kidneys in twenty were
enormously hypertrophied and the parenchyma
and cortical substance were so softened that the
finger could be easily pushed into the mass. The
other, which I found frequently, was defibrinated
blood and granules of black vomit extravasated
into the cellular tissue along the track of the
larger vessels of the vascular system and from

thètic. These two lesions may be considered
a part of the universal decomposition going on
in the organism.

But in my opinion the cadaver is not the right
place to seek for diagnostic symptoms; and
it is extremely doubtful if the pathological
anatomist ever lived who became so ex-
pert and familiar with all pathological lesions
occurring in the body, from any cause, that he
could by a careful and complete postmortem examination, and without any prior knowledge or
suspicion of the causes of death, pick out in a
dozen or more cadavers those that were known
to have died of yellow fever.

The living subject should be the field for re-
search, and the blood should be the special part
investigated; and this investigation should be
begun the moment it is known that the enemy
has gained an entrance. The blood is the
point of invasion, the material which constitutes
both the food and the tomb of the invisible,
mysterious and deadly foe. All the immediate
effects the irregular or paralyzed nerve action,
organic or functional derangements, and all arrest of vital metabolic processes are secondary,
and are being continuously augmented by the reflexes from the irritated and rapidly weakening
nerve centers, all of which reflexes result from the
destructive changes produced in the live stream.
This rapid destruction of the red blood corpus-
cels is followed by arrest of the functions of the

liver and kidneys; and on account of this impaired or suppressed elimination and the consequent accumulation of bile and excrementitious constit-uents of the urine, all of which add to and increase the disorder, distress and suffering reign in the system.

Doubtless every practitioner who has had much experience with this disease has seen, as I have, patients die in from twenty-four to thirty-six hours from the commencement of the disease; then am I not justified in saying that I believe that the poison begins to act as soon as it gains an entrance, and that this work of destruction is continuous and as rapid as that of the venom of the rattlesnake? The frequent analysis of the blood during the first twenty-four hours of the attack furnishes the best diagnostic points found in the *ensemble* of pathognomonic signs by which yellow fever is recognized; but, as I have said before, there is no one single distinct symptom by which it can be diagnosed at a glance.

It requires a close observation for several hours by a practiced diagnostician to make a positive and unerring diagnosis, although it is a specific and unique disease. But the same difficulty is found prevailing in nearly all other specific diseases; for who can sit down by the bedside of a patient suffering with a high grade of fever, on the first, second or even on the third day of its continuance, and diagnose it

"Variola!," having no knowledge that his patient
has been exposed to the contagion of that disease? No one; even in this enlightened day.
Still I believe that such a sign taken in connect-
tion with the assemblage of known symptoms
does exist not only in variola but also in yellow
fever and all specific diseases, as has been found
in pertussis and rubeola; when I said there was
none in yellow fever, I meant none had been
discovered as yet.

But suppose there was one well-known, clear,
distinct, and infallible symptom that would enable us to make a correct diagnosis at once; *cui
bono*? Has the physician discovered a balm
that will cure? Has he formulated a therapeutic measure or compounded an elixir that will in
the smallest degree modify, shorten or abort
this terrible disease? Has the profession, as a
whole, advanced a single iota in treatment since
the days of the great Dr. Rush with his lancet
and huge doses of calomel and quinine? Alas!
far from it; and now at this period of life, having
retired from active practice and being simply a
looker-on in the Venice of medicine, though one
intensely interested in all that pertains to improvement in the profession in which I have
labored so long, the conclusion is forced upon
me that the profession, by over-medication,
often prolongs a case by adding drug-disease to
the original one and thus frequently prevents the
recovery of some who otherwise would get well.

The experience and knowledge gained in later years convinces me that when in my first
years of practice I gave, as I had been taught,
ten to twenty grains of calomel, frequently adding twenty grains of quinine as an initial dose
in the treatment of yellow fever, I had aided the
enemy in destroying its victim; but I am thank-
ful to say that I earned, by my conservative
course afterwards, the equivocal compliment
from a wealthy client who said, in advising others to employ me, that if I did no good I was
sure to do no harm.

But to close the subject of symptoms and diagnosis I will say: Yellow fever is more easily
recognized in the first stage than is any other
one of the special diseases. When a physician
who is perfectly familiar with bilious fever, un-
expectedly and with no suspicion of its nature,
meets his first case of yellow fever, he is immediately impressed with the conviction that it is
something different from what he has ever seen
before—a something in the *tout ensemble* of the
facies, indescribable by words, which impresses
itself on the close observer in a way never to be
forgotten, and furnishes him an invaluable aid
to a quick diagnosis in the future. It is the *facies vultuosa*, caused by stasis of the blood in
the capillaries, a consequence of the primary
weakening effect of the poison upon the heart.

This stasis produces a reddish, florid color,
inclining to purplish upon the forehead, face and

extending down over the neck and the upper
part of the chest, afterwards changing to a bronze
tint, or to an orange-yellow color if general jaundice is to ensue. In the eyes, which are red-
dened and turgid, sometimes blood-shot, are
to be seen at first expressions of apprehension,
anxiety and inquiry soon changing, in serious and
fatal cases, to those of apathy, wild, fierce de-
lirium, or inexpressible horror, depending upon
the temperaments of the patients. There is
generally severe headache and much pain in the
back and limbs, attended with great restlessness
and nervous agitation, all due to and produced
by the rapid accumulation of bile and the con-
stituents of the urine in the blood.

The pulse is the next noticeable symptom
which will arrest the attention of the investigator. It differs greatly in these two fevers which
for two centuries were erroneously regarded as
identical in nature, only differing in degree, and
arising from the same cause. This also furnishes additional aid in diagnosis; for in bilious
fever, during the first twelve to eighteen hours,
it is full and frequent—one hundred and twenty
to one hundred and thirty per minute—with increased tension and force; while in yellow fever
the pulse reaches its acme, never over one
hundred and twenty, in the first three to six
hours; and the physician, at his first visit, generally finds it between ninety and one hundred,
soft and full, with but little force or tension; and

also finds a remarkable discrepancy and want of co-ordination between it and the temperature. The time of the onset of yellow fever is at night, almost invariably between nine o'clock at night and three in the morning; while no type of paludal fever, according to my observation, ever begins at that time of night. This fact and the course pursued by yellow fever, the most unique yet regular of any known disease, both afford important and valuable help in diagnosis. In fact, each and every phenomenon in the assemblage constituting the disease may be truthfully said to be pathognomonic of it; and the physician who familiarizes himself with them singly or as they appear in the group becomes as nearly infallible in diagnosing this disease as it is possible for man to be. This is no exaggeration. He can, by familiarity with the symptoms, as readily distinguish yellow fever from dengue and bilious fever as he can distinguish, by color and other characteristics, the pure blooded African from the fairest Castilian who ever vailed her unsurpassed lovliness from the vulgar gaze.

Having spent my long life wholly in paludal regions, and in localities frequently visited by yellow fever, it was but natural that I should have made these two most important fevers the subjects of special study and investigation.

But when I first began my investigations, my knowledge and ideas of these diseases were too limited, vague and crude to be of any satisfac-

tion to myself or real benefit to others; still I
persevered, under many difficulties and discour-
agements, determined to acquire all the knowledge that had been or could be acquired about
them; till at last I became not only very successful in treating both, but could also tell my patient the day and the hour when the malarial fever
would return as surely as the sun would rise that
day, if they did not carry out the prophylactic
course I had prescribed, but which nine-tenths
were very sure not to do because they were feeling so well that they thought it useless to continue taking medicine; and I have heard patients boast an hour before the paroxysm did
return that they felt better than they had felt for
a month.

This is very common in malarial fevers of all
types; and I infer from it that the cause of these
fevers produces first a stage of stimulation, by
irritating certain nerve centers, followed by a
reaction in the form of a chill and fever, constituting the stage of depression or probably
better a true neurasthenia. But while learning
this I found that nothing of the kind occurred
in yellow fever and that there did not exist
the faintest or most remote trace of relationship between it and this class of fevers, or
any other for that matter; but that it was so exclusively *sui generis* that it would never asso-
ciate or co-exist in the system with them, or
with any other acute disease. Thus I soon

learned to differentiate and easily distinguish it from any other fever or morbid condition, and I have been so fortunate in this regard as never since to have been mistaken in my diagnosis.

God knows I write this in no spirit of vainglory or boasting; for the consciousness and memory of the many errors and mistakes of my early years are ever present with me, and are too humilating ever to permit me to become vain or puffed up by all the knowledge that I can acquire of this or any other disease.

It has been my duty on three different occasions to be the first to announce the existence of yellow fever in interior towns, from the fact that the first cases among the refugees incidentally came under my care and treatment; and in every instance I had the misfortune to be at first opposed in my diagnosis and announcement by the rest of the resident profession; and in the last instance I was opposed not only by resident doctors but also by a dozen or more physicians and yellow-fever experts from other towns and cities. In the excitement and terror of last year (1897) I took a decided stand at the beginning, vehemently combatting the diagnosis of the official experts and contended that only dengue, in a more serious and grave form than usual, was prevailing not only in this state (Texas) but throughout the whole Southwest. Time, which proves all things, fully sustained and justified me in my diagnosis in the three instances first named; for a serious epidemic ensued in each case, and a number of valuable lives were lost which would have been saved by flight but for the number and unanimity of opinion of those dissenting from my report that it was yellow fever. As regards the epidemic of last year the evidence, at this time, is overwhelmingly in my favor that not a single case of yellow fever occurred in Texas, and is sufficient also to cause grave doubts of there having been any true yellow fever east of the Mississippi during the same period.

Having expressed opinions of the incompatibility of yellow fever with the paludal fevers and all acute maladies, opinions which are diametrically opposed to the teachings of two of our most learned and distinguished authors on the subject, Drs. LaRoche and Fagét, I deem it incumbent upon me, before taking up the subject of treatment, to give my reasons therefor and the evidence upon which they are based. But I will first say that I am inclined to think that the positions which these distinguished gentlemen occupied in the profession were too high to allow them time or opportunity to make such close, critical and scrutinizing observations (or to follow individual cases to their termination) as a physician can who occupies an humbler position and is not hampered by official or public duties. I think I am justified in expressing such an opinion from the number of oversights

and mistakes which I have witnessed in physicians filling professorships, and whom I regarded as much my superiors in their knowledge
of this disease. In every epidemic of yellow
fever, with which I have had experience, I have
witnessed the sudden cessation of malarial
fevers which had been very rife just prior to the
introduction of the exotic and major poison;
and the only cases of it with which I met during
the prevalence of the epidemic were in persons
who had been subject to recurring attacks of
the intermittent from every fourteen or twenty-
one days, and these persons were invariably
among the last to be attacked by the yellow
fever. During such an epidemic I have never
met with a case of malarial fever occurring
de novo in a person not previously subject to it,
except in Memphis in 1878.

Now I know that I shall be charged with
calling any and every case of fever occurring
during an epidemic, yellow fever. But correct
diagnosis has been my one special pride and strong
point during all the years of my practice; and I
do not feel that I have ever been guilty of the
mistake of diagnosing and calling a case of
malarial fever, yellow fever, or *vice versa*; for it
was between these fevers especially that I had
an ambition to learn how to make a differential
diagnosis, and to this end I labored and strove
during all the early years of my professional life.
I willingly admit that the two diseases are occa-

sionally seen prevailing in the same place at
the same time; but my observation has been
that the paludal fever gives place gradually and
completely to the specific one, and this spares
no one (except the pure-blooded African) from
the infant at the breast to the aged pilgrim
tottering on the verge of the grave. Hence I
think that the distinguished Professor Fagét
strained a point unsuccessfully in attempting to
prove that the children of Creoles were not liable
to yellow fever, when he endeavored to show
that the large number of cases of fever among
them, attended with hemorrhages and black
vomit, in the epidemic of 1853, '58, '67, were not
yellow fever, but "haematemesic paludal fever"
which prevailed epidemically those years con-
jointly with yellow fever.

It is passing strange that this peculiar and
rare type of malarial fever, which is only occasionally seen in sporadic form, should have
prevailed epidemically during the severe yellow-
ever epidemics of those years and never, at
least to any extent, before or since. I have
seen a dozen cases in one season since, but did
not regard this as an epidemic. It has never fallen
to my lot to see a case in which beyond all doubt
the two diseases co-existed in the system. As
I have said before all persons subject to recurring attacks of intermittent fever were among
the last to succumb to yellow fever, and then
not until the intermittent had been arrested.

And my observation has invariably been that in all cases where the paludal poison had possession, by preemption or priority of entrance, the yellow fever poison either could not or would not enter until the former had been dislodged or had thoroughly evacuated the premises.

Let me illustrate my idea further by a few cases. In Calvert, Texas, in 1873, I treated a young man in July, August and September for a recurring intermittent and failed to arrest its regular return every twenty-one days, because he was too busy to take the prophylactic remedies I had prescribed. In October my friend, the late Dr. Greenville Dowell, who was daily visiting myself and family during our attack of yellow fever, was called to see this same young man. His favorite remedies for yellow fever, calomel and quinine, quickly cured him; and the good Doctor, in his efforts to persuade me to take the same, related this among other cases to show the great success of his treatment. Now the truth was, this was simply the fourth recurrence of his intermittent fever without any sign of yellow fever at all; for late in November, ten days after safely passing the time for the fifth period, when he was congratulating himself on at last being rid of his chills, he was seized with a violent attack of yellow fever accompanied by black vomit, and died in a few days.

I met with a dozen or more just such cases in Memphis in 1878. I made it my special duty to

watch them all closely; and while three of them entirely escaped yellow fever, the rest were attacked
during November; but not in a single instance
did the yellow fever supervene until it had been
demonstrated that the intermittent had been
arrested, which was known by the patient's
safely passing an expected period ten days.

My driver, a young man aged 22, was a notable case. He regularly had an attack of intermittent fever every fourteen days during August,
September and October; refusing to take anything to break up or prevent the return of the
paroxysms because he had heard that a person
could not have yellow fever as long as he had
intermittent. So he lost two or three days every
two weeks, but he would get a younger brother
to drive for me until he was able to resume his
place. I treated his father and mother and
seven brothers and sisters, sick with yellow
fever during September and October, without
the loss of one. He escaped the intermittent
paroxysm due November first, and on the tenth
of that month he was seized with a violent attack
of yellow fever and came near dying; but he
finally recovered after six weeks' confinement to
the bed where I left him, the epidemic being
over.

But the saddest case remains to be told. On
October seventeenth, while eating supper at the
Peabody Hotel where the volunteers were all
lodged by the Howard Association, I received a

message from a very prominent volunteer physician, from one of the Northern States, request-
ing me to come to his room as soon as I had
finished supper. I knew he was subject to
recurring attacks of intermittent fever; and as
he had informed me that he had yellow fever in
Mobile more than twenty years before, I supposed there was nothing serious the matter.
Upon entering his room I found him sitting on
the side of the bed and clad in his underclothes.
As he had a rather dejected or apathetic expression of countenance, I greeted him pleasantly
with, "Well, Doctor, are you having another
one of those annoying paroxysms of malarial fevers?" "Yes," he replied, "but a very
anomalous one; for it postponed a week,
and I was feeling better than usual, when
to my surprise at three o'clock yesterday morning I had a very severe chill followed by a fever
lasting till noon today. I took three compound
cathartic pills last night and have taken about
forty grains of quinine in ten-grain doses today,
but I am feeling very queer tonight." I was
feeling his pulse while he was saying this, and
watching his countenance. To my intense surprise, I felt the characteristic yellow-fever pulse
and saw plainly depicted the unmistakable facies
of that terrible disease, exhibiting what I consider its fatal aspect.

With watch in hand I said: "Doctor, I have
an engagement to meet our medical director,

Dr. Mitchell, in five minutes, which I had forgotten till this moment. If you will excuse me I will run down to headquarters, see him and then ask him to come back with me to see you. I will not be gone ten minutes. Let me suggest that you lie down and cover up until I return." The Howard Medical Headquarters were only half a block distant. I found Dr. Mitchell, and said: "One of your volunteer physicians is quite ill at The Peabody, and I want you to go with me at once to see him." On the way he asked a number of questions; but I said, "Wait until you see him, for I want your opinion unbiased by mine."

Again the patient was found sitting upon the side of the bed. I saw Dr. Mitchell's eyes dilate as they fell upon him; and greeting him kindly by name, he asked abruptly: "Dr. K., when did your kidneys act? How much urine have you passed today?" The question seemed to confuse and irritate the Doctor, and he replied a little petulently: "Now Doctor Mitchell, don't think I have yellow fever, for I had it twenty odd years ago in Mobile, Alabama; but really, I do not remember having passed any urine today, or since yesterday afternoon for that matter." "Will you please allow me to try the catheter?" asked Dr. M.; consent being readily given the instrument was quickly introduced but only half an-ounce of urine was obtained. Withdrawing and replacing the instrument in its case and feeling his pulse, Dr. Mitchell arose saying: "Well, Doctor, you do not need anything tonight except a diuretic mixture and a good nurse to stay with you, both of which we will send you."

"But I do not need a nurse," replied the papient. "But," said I, "you may need to send for one of us during the night, or to send a message to some friend, so let the nurse come and stay."

There was no need to ask Dr. Mitchell if he thought as I did and saw what I had seen. Going down the stairway his only remark was: "How sad! Is it our duty to inform that doomed man that he is dying?" I replied that my experience was that no good ever resulted from so doing. Our unfortunate confrere died between three and four o'clock the next morning, forty-eight hours from the inception of the disease.

Many similar cases might be cited from every yellow fever epidemic prevailing in paludal regions; but the foregoing ought to be amply sufficient to clearly illustrate my idea and position as to the impossibility of the co-existence of the two diseases in the same system.

During this same Memphis epidemic, however, I met with certain peculiar cases illustrating the fact that yellow fever *can co-exist* with certain acute diseases, although, as I have shown, it cannot with intermittent fever.

On September twenty-fifth Dr. Mitchell said

to me: "I want you to take charge of Major
R., aged about sixty years, a particular friend of
mine, whose family, except his oldest son, is
out of town. I have been treating him for se v
eral days for a severe attack of facial erysipelas
which began on the side of the nose and now
involves the whole face. His face is so œdema-
ous that his eyes are closed, the line of demarka-
tion being just above the eyebrows. At three
o'clock this morning he had a chill which was
the beginning of an attack of yellow fever; and
as you are hunting unique cases I turn him over
to you, having confidence that you, being a
stranger to him, can treat him more successfully
than I can. I want ycu to do your best." In
twenty-four hours from the inception of the yellow fever all signs of erysipelas had disappeared
except a slight desqamation, and the case pro-
gressed so favorably, in every respect, that the
patient was able to leave his bed and sit up a
short time the morning of the ninth day. That
evening the erysipelas returned, beginning at
the line above the eyebrows (the point it had
reached when arrested by the yellow fever), and
continued its course over the scalp down to the
nape of the neck before it could again be arrested.

His son, a young man of thirty-two years, was
his principal nurse, and I cautioned him not to
inhale his father's breath while handling him. On
the first of October the son exhibited symptoms –

indicating that erysipelas was commencing in his throat. But he had a chill that night ushering in an attack of yellow fever, which ran a course very similar to his father's case. At my after- noon call, October ninth, I found him sitting up at a window, enjoying the gentle Southern breeze. I told him it was very imprudent, as I thought he still had to suffer with the erysipelas which certainly was commencing when he was taken with yellow fever. He went back to bed, saying his throat was feeling a little sore. At nine that night he had a severe rigor; but he did not send for me till six o'clock the next morning, and it was eight o'clock before I reached his bedside. I found him in a horrible condition. He took my tablet and wrote: "For God's sake give me something to put me out of my misery." His tongue was black and swol- len until it protruded between his lips so that he could not articulate. I sent a messenger for Dr. Mitchell, who arrived in a short time, and we did all in our power to relieve him, but in vain. He died a horrible death at noon the next day.

I will cite one more case to illustrate how gen- erally in an epidemic of that disease everything s called yellow fever, not only by ordinary phy- sicians but also by those occupying high posi- tions. By permission of my friend, Dr. Green- ville Dowell, who had charge of the Market Street Hospital, and at the patient's request, I

was called to treat Dr. S. H. McCormick of Saline City, Indiana, who was sick in the hospital.
This hospital was a four-story building, with a
capacity for accommodating two hundred patients, by crowding a little. Dr. McCormick
was in the fourth story and was the fifteenth patient in a rather small ward.

On the third morning Dr. Dowell invited me
to accompany him on his rounds through all the
wards of the hospital, an invitation which I
gladly accepted. He remarked as we started
that ninety per cent of all the patients who had
been brought into that building so far had been
sent out through the dead house. What a terrible report! But it was explained by the fact
that many were picked up on the streets and
brought in from elsewhere in a moribund condition. The Doctor's progress was so rapid that
it was with difficulty that I kept up with him,
and of course could derive no benefit from the
trip. But the Doctor, notebook in hand, prescribed for each case; but as the treatment was
routine, that was not difficult. At last, on the
third floor, in passing cot after cot, he remarked
as he passed a certain one: "This man is from
your town in Texas." Upon recognizing him I
asked why he was there. The Doctor replied
that he was brought in that morning sick with
yellow fever. "But come back here, Dr. Dowell," I said, "this man has had yellow fever in
a previous epidemic and cannot have it now."

"Oh yes," replied the doctor, "he has a genuine case."

I sat down by the cot, and taking hold of the man's wrist found a pulse indicative of an inflammatory fever. Calling him by name, I said: "Look here! Tell me the truth! What has happened to you?"

"Well, Doctor, in getting out of the carriage a few days ago, at a funeral, the horse started suddenly and I was thrown violently against the wheel hurting my left testicle severely."

"Yes, Joe," said I, "I heard that cart-wheel story in Philadelphia, in my student days, years ago. Let us see the amount of the damage."

Examination revealed an intense orchitis, the cause of his fever; and I said, "You had better get out of here, for nearly everyone who comes in dies." I lanced his testicle twice after he returned to Texas. He said he never was so relieved in his life as when he heard me say that he did not have yellow fever.

Upon arriving in Memphis, in 1878, I was much astonished to find such a virulent epidemic prevailing; and although familiar with it, the odor was so intolerable at the hotel that I requested the clerk to give me a room in the fourth story. After entering upon my duties in the district assigned me, I would dismiss my buggy in the suburbs at six in the afternoon, spending the night at the residence of some important patient who was always glad to have me

remain, and would order my driver to return in
time to take me to the hotel for breakfast.

I was also astonished to see the number of
inexperienced officers in this terrible battle—
young physicians, principally from the great
Northwest, who had never seen the disease and
who were totally ignorant of its nature. It occurred to me that this was a splendid opportunity to obtain statistics and make certain observations in regard to the disease, which I had
long desired to do. I therefore made it a point
to obtain the name, address and time of arrival
in Memphis of every one I saw wearing a Howard Medical badge. One evening after supper
quite a number of them gathered around me,
while I was entering data obtained that day, and
were very curious to know why I was thus getting their names and addresses. I replied:

"Young gentlemen, I am deeply interested in
the study of yellow fever of which I know nothing, though I have been seeing it at intervals for
twenty-five years. I am also much pained to see
you manifest such careless indifference to the
great danger into which you have unwittingly
rushed; for let me tell you that in the fiercest
battle of the late war the danger to the same
number of officers was never half as great as
that to which you are all exposed." They laugh-
ingly replied:

"Why, it is only a malignant form of bilious
fever, totally non-contagious, and we are going

to cure it with plenty of calomel and quinine,
and keep it off by taking quinine." To this I
said:

"If one in twenty of you escape an attack I
will be much surprised"; and showing them
my book in which I had entered their names; n l
drawn lines for columns headed, "date of arrival, date of attack, period of incubation, result,
recovery, died," I added, "and I very much fear
that many of your names will appear in the last
column."

I obtained a list of forty-five, to which Dr.
Mitchell added nine more. Of my forty-five
only one escaped having this fever, one who
wisely deserted and returned home after that
night's talk; and thirty were buried in the beau-
tiful cemetery of Elmwood in less than thirty
days from that hour. Many of them requested
me to attend them, if they should accidentally
contract the disease; but the confusion that
reigned and certain uncontrolable circumstances
combined to prevent this, so that I only attended
three, one of whom being too obstinate to listen
to advice from any one died in the second relapse. The shortest period of incubation was
one day, the longest twenty-five days; and in
these two instances also the patients, for weeks
previously, had suffered recurring attacks of intermittent fever. Noble band of martyrs! In their
anxiety and effort to relieve the distress and suffering surrounding them, they ignored their

own danger and died without a thought of self.

Upon the announcement of the first cases in July, 1879, I was appointed inspector under the National Board of Health and ordered to Memphis to investigate their origin. This appointment was due, I suppose, to two members of the board, Doctors Mitchell and Bemiss, who well knew my theory of hibernating germs. Upon arriving in Memphis about July twelfth, I found that on the ninth three cases had occurred in different parts of the city a mile apart and apparently having no connection with each other. There had been a celebration of the Fourth on the bluff, at which there was an immense gathering, and the day was intensely sultry. I found that certain parties had ordered two box-cars of bananas from New Orleans especially for this celebration; and after three weeks of patient investigation and inquiry, I established the fact that the three cases first reported were in persons connected with the handling and sale of those bananas.

Knowing from past observation that this particular fruit was a splendid fomite for the transportation of yellow-fever germs, I obtained from the records of the railroad office the numbers of the box cars in which this fruit was brought to Memphis and the date of their arrival. Armed with this information, I went to New Orleans; and going carefully over the Custom House records of the arrival of every vessel from the first of May to the thirty-first of July, I selected the steamer E. B. Ward, Jr., which had arrived from Blue Fields, June twenty-seventh, as the vessel most likely to have brought in those bananas which I believed to be the cause of the outbreak.

Further investigation developed the fact that the two box cars in question had been run down alongside of the Ward, June twenty-eighth, and loaded directly from her hold with bananas for Memphis. This steamer belonged to that wealthy Italian who a few years afterwards, with a number of others, was shot to death by a mob in the jail of New Orleans; and was one of the line of vessels running to Blue Fields under special privileges, not being inspected or detained at quarantine.

When I announced, as the result of my investigations, that this vessel had brought the germs of yellow fever that was then prevailing in New Orleans and Memphis, a vigorous protest was raised; and a number of prominent business men and managing officials of transportation lines said:

"Doctor, it will never do to make such a report. Why, not a single case of yellow fever has occurred on that vessel this season!"

"And a good reason why there has not," said I, "for her astute owners have selected her crew with that object especially in view; all are well-known immunes."

Two of the crew informed me that the steamer called at Havana on the return trip of that date; but from lack of authority and opportunity I failed to obtain their sworn statements to that effect. So once more commercial and pecuniary interests were powerful enough to suppress truth in reference to this terrible scourge.

And now what shall I say upon the subject of treatment, which really does not come within the purview of this work and the discussion of which properly belongs to those who are still in active practice and yearly seeing the disease? Although I have been deeply interested and have spent much time in the study and investigation of the origin, cause and nature of yellow fever, still I have never lost sight of and have tried to keep pace with all the novelties and supposed modern improvements in its treatment. Nor have I found it a difficult task; for, after an impartial review of the subject and a study of the different methods of treatment that have prevailed from its earliest history to the present time, I must candidly say that I fail to see wherein classic medicine has made any advance or improvement upon the treatment of Rush and Physick, barring their lancets, or can boast of any better success; while homeopaths, expectants and other irregulars have jogged along, *pari passu*, claiming equal if not better success and that too not without good show of reason and statistics.

At that last meeting of the Howard Medical Association of Memphis, reports were made by different members of the corps of the individual methods of treatment, statistics of success, etc., during that terrible epidemic of 1878 then rapidly drawing to a close. While the treatment of no two of them was exactly alike, yet when analyzed the differences were found to be immaterial; for calomel and quinine seemed to initiate and calomel and quinine to close it. That was active medication indeed; for death frequently dropped the curtain on the closing scene unknown to the physician, as I accidentally learned. Among those reporting were some on my list who had passed through an attack successfully, and were rather exultant over their method of curing the disease with teaspoonful doses (as they expressed it) of quinine. But unfortunately they had not learned the course of the disease, even from their personal experiences, and supposed that the subsidence of the fever with the deceptive calm at the end of seventy-two hours was the result of their large doses of quinine and the end of the disease; and on account of the number of calls which they had to attend they dismissed such cases from their minds, thinking that they were really convalescent.

After our active labors were over, our Medical Director requested me to make a careful canvass of several of the districts for the purpose of

obtaining certain statistical data as well as to
learn all the facts surrounding the death of several of the volunteers on my list, who had died
in private houses, that we might be enabled to
answer intelligently letters of inquiry from their
families. It was while I was engaged in this duty
that I learned the facts just stated; and I was
pained to hear, in many houses, not loud but
deep and bitter curses against the physicians for
giving such large doses of quinine to their loved
ones and then heartlessly, as they supposed,
neglecting them. I also learned that, still la-
boring under the erroneous idea of a relationship between malarial and yellow fever, they had
advised a continuation of these doses three
times a day after the subsidence of the fever to
prevent its return! So in my last paper before
the Association I thought there could be no im-
propriety, considering the difference in our ages,
for me to say: "If those gentlemen who have
been advocating the use of quinine in such large
doses would take the trouble to go over their
part of the battle ground again with a list in
hand of the patients treated, they would be
astonished to find the number of those they had
marked convalescent who had since died."

Like all young graduates I had reverence and
love for my Alma Mater, and implicit confidence
in the wisdom, knowledge and infallibility of
those whose names were appended to my di-
ploma. But this glamour quickly wore off under

the effects of personal observation and stern bedside experience; and I must confess that I was
not long faithful to the *magister dixit;* especially
in reference to this disease. I soon found that
personal observation and experience are the only
masters who never deceive and upon whose dicta
we can always depend. Under the guidance of
these teachers it will not take the observer long
to find out that yellow fever will run its peculiar but regular course unaffected, whether medicine has been administered or not. Upon learning this fact, the question immediately presented
itself: What then is the indication for the ten
to twenty grains of calomel and twenty grains
of quinine with which I was taught to begin the
treatment?

Further experience and observation taught me
that the condition produced in the system by the
effects of the poison was one not of sthenia or
inflammation, as believed and taught by the
great Doctor Rush (which belief caused him to
use the lancet), but one of asthenia, general debility and lowered vitality. Hence not only the
utter uselessness but the extreme harmfulness
of these two potent remedies which I had used
so freely, for their effects tended only to increase
the existing debility and still further to lower vital
energy. I saw that medicine had discovered no
antidote; nor had it instituted any method of
treatment that could modify, shorten or abort
this terrible disease. And while I regarded ex-

pectancy as pusillanimous in the physician pro-
fessing ability to combat disease, yet in the light
of my experience so far I considered it prefer-
able to a course of active medication which I
had found to do more harm than good.

I further learned from observation and actual
bedside experience that a large per cent of the
whole number of cases in an epidemic would
recover without any medication whatever, if they
were controlled properly, nursed carefully and
kept from exposing themselves or committing
any-great imprudence; while on the other hand
I found that a smaller per cent, although constituting a fearful rate of mortality (say twenty-
five per cent of the whole number), would die
in spite of all that medical art had learned to do
for them. I believed it to be the observant physician's first duty, and not a difficult one either,
to learn in every epidemic to distinguish quickly
between those non-fatal and fatal cases, so that
he might avoid increasing the rate of mortality
by too much officiousness and over-medication
in cases that would naturally recover—faults unfortunately too common in young physicians
during their first experience with this disease.
I do not hesitate to admit that such was the case
with myself; and as I have been a close observer
during my whole professional life I have seen the
same faults in others.

Of the one hundred and eleven volunteers on
the Howard roll at Memphis, in 1878, how

many could claim to be cool, close, critical observers, especially in those first hours of confusion and terror? Less than fifty of them had
ever seen the disease before; and the impression made upon me, as I watched them start
eagerly for the districts assigned them by our
Director General, was that there was a braver
set of men than any that ever charged a battery
upon the most fiercely contested battle field in
the world's history. Ay! braver than Tenny-
son's "six hundred" as they rode, on that mem-
orable occasion of folly, "into the jaws of death."
A poet laureate sang their brave but useless
deeds, and a grateful nation immortalized them
in brass and marble; but the nobler deeds of
these unselfish heroes have never been sung. Day
after day and week after week they unfalter-
ingly met and battled with that unseen foe,
passing through scenes in comparison with
which the most hideous of battles pale into utter
insignificance. War furnishes scenes of mangled
men and animals only, while those resulting
from the murderous work of this King of Ter-
rors often included whole families — father, mother
and children (four, five, six, and seven) all dead
in one room, and lying in every conceivable
posture. In one instance a tender babe was the
only one found alive, and it was vainly endeavoring to draw nourishment from the breast of its
mother lying dead upon the floor; in another,
the dead babe was found with its little gums

clinched around the nipple of the left breast of its mother who was sitting stiff, stark dead in a chair.

Such were some of the scenes which met the eyes of many of these physicians in their first rounds; and although some of the few survivors said to me that they felt their hearts blanch as they had never done when in the deadly breach of battle through which they had passed, blanch with an undefinable fear or momentary dread that certain annihilation awaited them as well as the doomed city, yet they neither faltered nor grew weary of their self-imposed tasks but were untiring in their efforts to stay the ravages of the plague. In fact, after the slight shock of the first reconter had passed they were seized with a fascination for this new, strange and hidden foe that they had volunteered to fight; and forgetting self, like the bird charmed by the rattle snake, they labored and fell (as many as six in one day) at their posts of duty till the rate of mortality among the non-immune volunteer physicians in Memphis reached sixty-six and two-thirds per cent! Two-thirds of that noble band of volunteers died on the doctors' battle field, under circumstances which should entitle them to wear the crown of martyrs as well as heroes. "Greater love hath no man than this, that he lay down his life for his friend."

These facts first attracted my attention and set me to thinking of this peculiar scourge as a

Nemesis, an agent of retributive justice for the punishment of a great sin which for commercial and pecuniary gain had been jointly carried on, for nearly two centuries, by all the civilized and Christianized nations of the earth, as a legitimate traffic (God save the mark) against the most benighted and ignorant race on the globe. Notwithstanding the eloquent sermon preached by Rev. H. C. Morrison, at the Broadway M. E. Church, South Louisville, Kentucky, Sunday, October 18, 1878, in which he took the position and endeavored to prove that yellow fever could not be in any sense a scourge or punishment for a sin which the Southern people could be held as guilty, I still held to my idea. But the reverend gentleman was plainly ignorant of the true history of its origin and especially of its intimate and constant connection with that nefarious traffic, the old African slave trade, all of which has been detailed in the preceding pages.

Now, to finish the subject of treatment: Having discarded the orthodox method I learned in the halls of classic medicine, I adopted the simplest method possible, not to be "expectant" though bordering closely on it. First: I cleared the bowels perfectly, as early as possible in the disease, with a dose of castor oil or a large enema of hot water, not disturbing them again till after the fever subsided. Second: I kept the

patient perfectly quiet in a horizontal position,
never allowing him to rise up even for a drink
of water, nor to see any one except his physician
and nurses; and to be perfectly on the safe side
this was kept up religiously for eight or nine
days (this is not absolutely necessary in all mild
cases).   Had this been done in Memphis many
valuable lives would have been saved.

Let me mention three notable cases to show
how easily and by what a little thing, in this disease, the scale is turned against the patient.
When Mr. R. A. Thompson, the postmaster, was
stricken, his friend, Herbert Landrum, the
quick, witty, sparkling, and bright city editor of
The Avalanche, took him to the home of his
father, the Reverend Doctor Landrum, pastor
of the Central Baptist Church, and tenderly
nursed and cared for him.   On the third day,
the day of the treacherous, deceptive calm, he
was free from fever and thought to be convalescent.   That morning he changed his pillow and
position from the head to the foot of the bed,
contrary to the advice of his physician, Dr.
Mitchell, made himself comfortable and read the
morning papers while he leisurely partook of
some tea and toast.   In sixteen hours he was
dead.

His friend, Herbert Landrum, was taken
down in a few days with a mild attack; and
although this example of the terrible effects of a
little imprudence was fresh in his memory he

got up on the fourth day and went to his office
to answer a few letters. In less than forty-
eight hours he too was dead.

Mr. Catron, a very close friend of both,
nursed and helped to bury his two friends; he
was mildly attacked in a few days, committed a
little imprudence after it was thought that he
was convalescent, and soon followed them to
the cemetery.

It will readily be seen from these cases that
good nursing and prudently remaining in bed a
sufficient time is worth more than all the medicine that can be given; and I mention them as a
warning, hoping that they will be the means of
saving valuable lives in the future.

First: Keep the head cool by bathing it with the
old French "*Eau Sedative*" and the feet warm by
applying bottles of hot water to them. If the
skin is hot and dry during the fever direct the
nurse to sponge the patient under cover every
two hours with tepid water and whiskey, wiping
thoroughly dry each time; but if the skin is
moist during the febrile stage this sponging will
be unnecessary. If there is much pain in the
back or the head, apply to the spinal column a
flannel strip saturated with spirits of turpentine;
put a dry cloth over this, and for ten or fifteen
minutes gently rub a flat iron, as warm as can
be comfortably borne, up and down the spinal
column. This is the most grateful thing you
can do for your patient and will not only

relieve the pains but at the same time will cause
the whole body to perspire gently.

In the olden time I frequently applied six or
eight cups along the spinal column, wet or dry
depending upon the condition of the patient;
and I will add just here in regard to this old
procedure, which seems to have been entirely
discarded of late years because there is a little
trouble connected with it, that when there is intolerable nausea or incoercible vomiting in yellow fever or any other malady wet cups applied
freely over the ganglia of the celiac plexus
and the œsophageal branches of the pneumo-
gastric, from whence nerves are distributed over
the stomach and other abdominal organs, will
relieve these intolerably miserable and sometimes dangerous conditions more promptly
and effectually than any internal or other
external remedy. Allow the patient all the
cracked ice he desires; use diluents freely, such
as teas made of watermelon seed or flax seed
and flavored with fresh lemon juice, hot or cold
to suit the patient's taste; but be careful never
to offend the stomach by forcing upon it anything that it dislikes.

No arterial sedatives or other febrifuges are
required or are even permissible; for the first
and principal effect of the poison is upon the
heart and the blood, and the pulsations are too
rapidly diminished from this effect to allow
agents to be used that will increase it. From the

inception of the disease strychnine and digitalin, in doses of half a milligram (gr. 1-134) each, should be given every hour to maintain nervo-vital energy and to sustain the heart's action.

The foregoing was my first departure from the teachings of the Masters, if departure it may be called. I have yet to find anything that does not belong to regular or classic medicine, or to find a court for the final decision of disagree-ments in matters of this kind. If this were not so there would have been no necessity of Pope's asking: "Who shall decide where doctors disagree?" But later, learning of M. Polli's discoveries in regard to the sulphites and bisulphites and their uses in medicine and especially in zymotic diseases, and being encouraged by my use of them in variola, rubeola and chronic paludal fevers, I determined to try them in yellow fever, especially the hypo-sulphite of soda which I knew would do no harm even if it accomplished no good.

Now I went to Memphis to do all the good I could for that sorely afflicted people, but also for the purpose of making a special study of certain points in the disease which I knew I could not do if I were over-burdened with patients. At my first meeting with the Howard Medical Society, which met at nine o'clock every night with Doctor Mitchell in the chair, to listen to the daily reports from this army of one hundred and eleven volunteers, I heard some of my new

friends, whose names I had been taking down
that very day, report having attended seventy-
five to one hundred cases. I thought it was my
time to turn pale, as I had only twenty to report; and I knew I was not physically able to
see and treat such a number of cases, scattered
as they were over the district—old Fort Picker-
ing and South Memphis—to which I had been assigned the day before.

I therefore inquired what they did for such a
number of patients. "Oh, we took down their
names and addresses, with street and number,
to report at this meeting." I replied that I
thought that that was the duty of the Howard
visitors, of whom the Association had quite a
number employed, as well as to see that they
(the sick) obtained their medicines promptly;
but that I found that the first prescription I
had made the morning before had not been filled
until that evening on account of the rush and
the number ahead of it. I said I thought we
had better all go home for the good we were
doing, unless Dr. Mitchell would permit us
to carry what few medicines we needed and
dispense them ourselves; for there were not
prescription clerks enough in the city to dispense
one hundred of the prescriptions made daily.
Moreover, my plan would mean a great saving
of expense to the Association; for the prescription which I had referred to (sodium hyposulphite, one-half ounce; syrup of orange, one

ounce; water, four ounces) was charged to the Association at the very moderate (?) price of two dollars, and was received by the patient too late to be of any particular benefit. Dr. Mitchell said the suggestion was a good one; and from that time on the physicians carried the necessary medicines the Association having ob- tained possession of a wholesale drug store that was closed and put in volunteer clerks who sup- plied us with whatever we needed.

I said moreover that I intended to see each of my patients twice a day, and therefore could not add a greater number of new cases daily than I discharged old ones. I treated about two hundred and forty patients in Memphis, of whom about thirty were malarial intermittents to whom I gave but little attention; but I selected one hundred typical cases of yellow fever, of which I kept a regular daily record of the pulse, temperature, tongue, bowels, urine, respiration, appetite and sleep. Of these I I treated fifty with sodium hyposulphite as the dominant medicament and the other fifty with sodium sulphocarbolate, giving these medicines in doses of twenty grains every two or three hours; and in cases where adynamia predom- inated I also gave ammonium phosphate in the same dose, and in extreme cases I gave ammo- nium carbonates in doses of five to ten grains.

Whether the treatment pursued in these cases had any beneficial effect I am unable to say; and although there were only twelve fatal cases among them, a reasonably small death rate, yet I am so skeptical about any good ever having been accomplished in the disease by any course of treatment with the old galenical preparations that this question arises in my mind: Would not the eighty-eight have recovered anyhow, if I had not seen them or if not a single dose of medicine had been administered to them, provided they had been kept in bed and properly nursed.

I had not then learned the method of "Modern or Improved Allopathy," taught by the celebrated professor of Ghent, Dr. Adolph Burggræve, although it had been known and used all over Europe for ten years or more. It is strange that the physicians of the United States, who are noted for being so progressive and up-to-date and even in the lead in many of the modern discoveries and improvements in the science of medicine and who eagerly received and tried Koch's Tuberculin, Brown-Sequard's Elixir of Life, etc., should have been so slow to investigate and try this very simple yet truly scientific method of exact, positive medication with the alkaloids and other active principles of plants. Its distinguished author, who deprecated its being called a new system of medicine, named it "Dosimetry" meaning measured doses. He and his co-laborator, M. Chàrles Chanteaud of Paris, devised machinery by which the alkaloids and other active principles were thoroughly triturated with sugar of milk and then mechanically divided into mathematically exact doses and shaped into granules, thus putting into the hands of the physician "arms of precision" upon which he could rely implicitly and by which he might secure with extreme exactness any physiological or therapeutical effect he desired.

This admirable method has been tried by a few Brazilian physicians in the treatment of yellow fever, though not in accordance with my ideas of the nature of the disease. But as I am convinced that no method of medication or any kind of medicine will have any effect upon the disease in those rapid cases which seem to be over-whelmed by the poison and stricken to die in a short time in spite of all that medical art can do for them, so I should not expect any better success from the use of this method except as a prophylactic measure. I have given dosimetry ten years of close, careful and unbiased investigation and trial in all the acute diseases which prevail in this southern country; and I have found that the most of them can be prevented, aborted, or jugulated by it—this of course depending upon the time treatment is instituted. Hence we may reasonably expect to accomplish the first result, prevention (which is always better than cure), by the use of this method during an epidemic of yellow fever.

There seems to be as much difference of opinion

regarding the nature of the disease as there is
regarding its cause; and this accounts for the
great number of therapeutic remedies that we
find advocated in its treatment. The majority
of writers claim that it is the nervous system
which is first attacked by the poison of the disease; and yet at the same time they inconsistently
classify it among the zymotic affections.

Now if this theory of its cause which I have
been advocating is true (and this is too clearly
and perfectly sustained by all known historical
facts concerning the disease for it ever to be successfully controverted) then it ought to be plain
to every intelligent investigator that the specific
infection causing yellow fever is a blood poison,
and that the toxicohemia is the result of a virus
or animal poison which introduced into the cir-
culating fluid produces septic or prutrescent
effects in the blood which is the point of invasion. This is strongly sustained by the fact
that putrefaction takes place more rapidly after
death from this disease than any other, and also
by the so-called black vomit which is far more
frequent than even experienced physicians are
aware of; for I do not recall a single postmortem that I ever made or witnessed in which
abundant evidence was not found of the commencement of this change or decomposition of
the blood, although in many no evidence of it
was seen during life. In fact, it seems to me
that it is a misnomer to speak of it as a vomit;

for the stomach is not its only source by any means; it oozes out through the mucous surfaces of the mouth, throat, œsophagus, stomach and intestinal canal, and also transudes through the weakened and distended walls of the blood vessels.

You who have had much experience with the disease are doubtless familar with the scene of the patient every now and then throwing his head back and spurting a mouthful of the horrid stuff against the wall, upon the floor, the bed or upon your person, or even into your face if you are not on your guard. This is what trickles down from the gums and fauces and does not come from the stomach at all; and it is not till there is an accumulation of it in the stomach that it is vomited, unless that organ is in an irritable condition and vomiting is occurring from other causes.

All my observations before and after death tend to confirm the opinion that it is the blood which is first invaded and affected by the specific poison of yellow fever; and remembering that the blood makes the round of the circulation in sixty seconds, or a little less, it can be understood at a glance how easily and quickly the nervous system as well as all the tissues and organs of the body, can become affected and all their functions deranged.

I have said that I should not expect any better success with the dosimetric method of treatment of yellow fever than with any other, except as a prophylactic measure. But I should not make such a sweeping declaration after the great success I have had with it during the last ten years of my active practice, for I have not had the opportunity of trying it in this disease. It would afford me great pleasure to have the opportunity of treating, according to this method as a curative measure, one hundred typical and serious cases of yellow fever after the disease had actually commenced, and also a hundred or more non-immunes by the same method as a preventive or prophylactic measure. I have seen all other infallible (?) prescriptions and vaunted prophylactics tried and fail; but I have perfect confidence that the dosimetric method would prove a success as a prophylactic at least; and I should not feel that I was experimenting with human life in trying this method in yellow fever for the first time, especially after having seen all other methods tried with but little benefit.

Moreover, this method does not discard any old prescription or remedy which has been tried and found useful. I said it was "improved allopathy;" it is more, it is an improved method of therapeutics. And while it is a very plain and simple method, yet the physician who would practice it successfully must become a good therapeutist; and to do this he must be well versed in every department of medicine, and also be a close observer and a good reasoner.

Besides this, whether it be the nervous system or the blood which is first attacked, this method points out exactly what is necessary to be done in either case.

As all schools admit that prevention is better than cure, I will, from my experience with this method as a prophylactic in other serious diseases, give the course I would advise all non-immunes to pursue—a course which I think best calculated to prevent the disease entirely, or failing in that to at least so modify it as to render it benign and innocuous. Take one granule of arseniate of strychnine, gr. 1-134, and two granules of arseniate (or hydroferrocyanate) of quinine, gr. 1-6 each, every one or two hours till six doses are taken in a day. At the same time take two granules of calcium sulphide, gr. 1-6, five or six times a day. Take no purgatives, and end each day at bed-time by taking two granules each of the celebrated dosimetric trinity—arseniate of strychnine, gr. 1-134; aconitine, gr. 1-134; and digitalin, gr. 1-67. This combination will help nature to restore the vital energy used up during the day, equalize the circulation and strengthen the heart's action, as well as procure quiet, restful sleep. Occasionally take, what Doctor Burggraeve calls a *lavage intestinale*, a teaspoonful of seidlitz salt—Abbott's Saline Laxative (neutral sulphate of magnesia in effervescent combination) in half a glassful of water, before breakfast, followed by sucking

the juice of half a fresh lemon. This will carry
off the previous night's accumulation of vitiated
matters and mucus from the mouth, stomach
and intestines, thus preventing foulness of the
stomach and constipation. This general prescription should be kept up for ten days after
entering a yellow-fever infected district, and
then omitted for a week; but, to be on the safe
side and take all necessary precautions, I would
advise non-immunes to take during this inter-
val two granules of strychnine hypophosphite,
gr. 1-134; and two or three tablets of Nuclein
(Aulde), two drops each, four times a day.
These will sustain vital energy and repair the
waste which is going on more rapidly than under ordinary circumstances, and will thus prevent the entrance of any pathogenic microbes
into the organism.

The foregoing is all that I would advise in the
way of medication; and it ought to be kept up
for the first month or six weeks of exposure, and
at intervals during the entire epidemic, depending upon the amount of labor and exposure and
fatigue that these parties have to undergo in the
discharge of their duties. And while there is
apparently a great deal of medicine involved in
the course, judging from the number of doses,
yet, when summed up, the aggregate will be
found to be an extremely small amount—say
about twenty grains of quinine and twenty grains
of calcium sulphide, with less than one grain of

strychnine, in ten days. It was this method, especially through the labors of Dr. Fontaine, of Barsur-Seine, that brought the sulphide of calcium so prominently into use as a remedy in all 'zymotic diseases, and which Dr. Burggraeve terms "the parasiticide par excellence," while he has named the arseniate of strychnine his *"cheval de bataille."*

I will attempt no outline of what my curative treatment of yellow fever would be according to this method, but will say that under any and all methods it requires no potent nerve depressors, no powerful arterial sedatives nor antiphlogistics of any kind; for it is from the beginning a terrible state of asthenia with a tendency to rapid destruction and putrefaction throughout the whole organism, and requires all the skill and all the agents of our art to counteract it and to sustain struggling nature with nervo-vital incitants, tonics and reconstructives until the danger is passed.

In speaking of the use of sulphide of calcium in all zymotic diseases I did not mean to imply thereby that yellow fever belongs to that class, though I have been guilty of so doing in the past simply from force of habit and because it has been so classified by nosologists generally. This, however, is an error; for it does not belong to that class at all. It is essentially a septic fever caused by the introduction of a specific, septic, animal poison into the circulatory system. thus producing a true blood poisoning.

As the difference between zymotic and septic poisons in their course, action and effects is seldom mentioned and as they are sometimes used synonymously, I will give you my idea of the difference between them—an idea derived from long comparative observation of the various diseases produced by them; for to my mind a clear comprehension of this difference greatly aids us in understanding several disputed points in regard to the nature of yellow fever.

The zymotic contagium—be that whatever it is, whether vegetable germ or zo-ospore—requires a nidus within the human body in which to incubate and then hatch. During this period of incubation no appreciable change or reaction follows but at the end of this period, which is longer or shorter according to whatever the disease is, the pathogenic virus is hatched out, and infects the whole economy at once, giving rise to general symptoms which are common to all diseases of the class. So that in diseases produced by zymotic poisoning there are three distinguishable stages: first, an incubative or dormant stage without symptoms; second, a stage of hatching and general infection accompanied by general symptoms common to all of that class; third, a stage in which the specific symptoms pathognomonic of that disease appear. In measles there is first the incubative stage lasting from ten to fourteen days; second, the ca-

tarrhal and febrile, three to four days; third, the
eruptive, three to five days.

There are no such observable stages in diseases produced by septic poisoning; for these
poisons incubate and hatch into the pathogenic
virus outside of the body, which then gains entrance into the body through the natural passages by being inhaled into the lungs with the
air breathed and swallowed into the stomach
with the liquids and foods taken by the individual in a locality infected by the virus. In the
case of a true germ disease, like yellow fever,
countless myriads of the pathogenic microbes
gain entrance thus far; but the vast majority
and in many cases all of them are destined to be
destroyed by the vital energy of the organism
into which they have gained entrance. But
when they once meet with a vulnerable surface
through which they can enter the blood-stream
their work of destruction begins and, in consequence of the rapidity of the circulation, reaction commences and the whole body is infected
at once and specific symptoms follow immediately.

The next general difference between these poisons is that in septic diseases the gravity, violence and result of the attack depend in each
case upon the virulence and quantity of the septic poison introduced; while in zymotic poisoning this makes absolutely no difference in the
violence of the reaction or the result.

Finally: In contagious, zymotic diseases the disease-producing germs or microbes continue to propagate and multiply in the organism which they have invaded and escape from that organism to infect others, thus causing new cases; whereas the septic poison reaches the limit of its existence in each case infected and is either vanquished and cast out as effete matter; or it overwhelms and destroys the individual and, Sampson-like pulling down the pillars of life's citadel, is involved in the ruin wrought by it. Hence, as a rule, septic diseases are not contagious, while zymotic diseases are.

This explanation of the principal points of difference between the zymotic and septic poisons enables one to understand easily why yellow fever is not contagious from the person; for the pathogenic bacilli, after accomplishing their work, perish in the body and thereby become innocuous; and as they do not multiply or propagate in the body there can be no infectious or contagious emanations from that body. It also enables one to clearly comprehend how and why a refugee from an infected district, although he may not have the disease himself and may even be an immune, is a source of danger to others if he comes to them in the same suit of clothes which he wore while in the infected district, and particularly when he brings a trunk of clothing packed in that district.

Thus my readers will be enabled to understand

why I advocate a quarantine; not against the person of the refugee, but- against his clothing and against everything he brings from the infected district which may serve as fomites for the transmission of the disease from which he is fleeing.

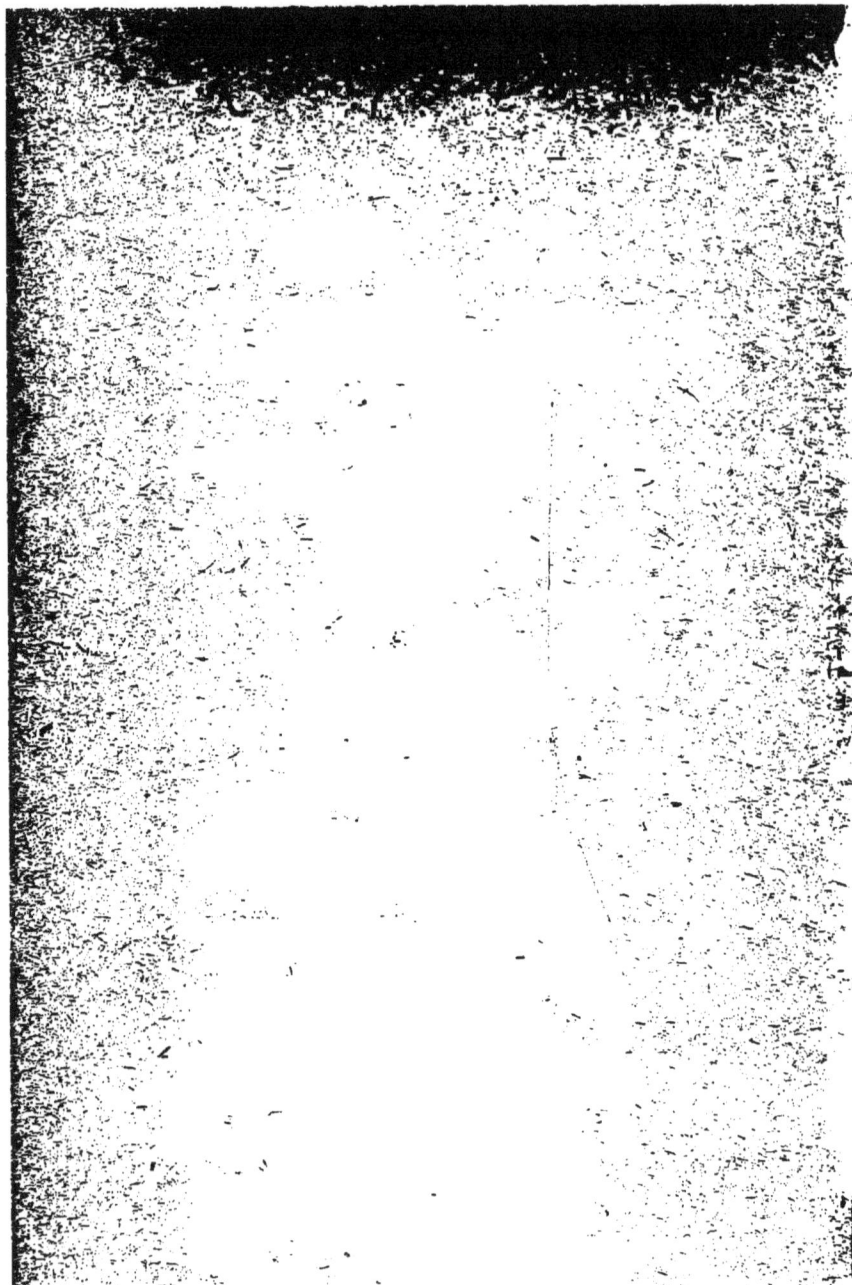

APPENDIX.

Containing a parallel table of the principal
characteristic symptoms of Yellow Fever and
Dengue, and so comparing them as to show how
a differential diagnosis can be made and the two
diseases distinguished. Also, a few more important facts and data which were unintention-
ally omitted, and which greatly strengthen the
theory of the origin of Yellow Fever and all the
positions and predictions concerning the disease,
closing the whole matter by a repetition of how
the disease may be eradicated.

APPENDIX.

There are a number of reasons which impress me with a sense of the propriety and even the necessity of closing this very imperfect history of yellow fever with a short chapter upon what is frequently termed its twin sister—dengue, a disease equally as unique and mysterious as the former, and as little understood by the profession.

Principal among these reasons are : First, the position I took in reference to the epidemic of 1897; second, since dengue is frequently taken for yellow fever, scarlet fever, and some other diseases, to show the great difference between the two diseases; and third, to acknowledge my error, if I was wrong last year, and to show how it was possible for me to have been so, thus convincing my readers that I am not in the least biased or governed by prejudice or bigotry but am desirous that truth alone should prevail even at the risk of my being misunderstood.

I took a firm stand at the beginning of the alarm, expressing the opinion that there was not and would not be a case of yellow fever in Texas this year (1897) and that there

APPENDIX.

od reason for doubting that the disease
revailing east of the Mississippi was yel-
ʒer. I have always been very decided in
gnosis, though sometimes slow to arrive at
t have always been open to conviction.
earlier years of my practice, during the
ɛnce of an epidemic of dengue, I once
sed a case as scarlatina and was very
ɛnt that I was correct until convinced of
ʒor by an older confrere and by subse-
events in the course of the epidemic.
I candidly confessed my mistake and
1 by it afterwards. I have since known
ʒer of physicians to make the same mis-
ʼho would not be convinced by all the
ɛnts and facts that could be brought to
pon the points. And I have been afraid
ɩe official experts of last year were a
nclined to be of that stubborn class
ɩought it beneath the dignity of their
ɑ to confess that it was possible for them

, although I was not in active practice at
e, I had been watching closely the reports
progress and course of the dengue (which
ɩs I can learn began in San Antonio) for
h or more prior to any excitement or talk
ow fever; and having passed through a
r of such epidemics and listened to the
ɛs and arguments of members of the pro-
as to what it was, I was prepared

to hear it called yellow fever, scarlatina or almost anything else calculated to alarm the people. To do what I could to prevent the frantic stampede which I knew was bound to follow the report of the official experts then making investigations in Galveston, I wrote several short articles for the daily paper of Navasota, Texas, where I was staying temporarily, advising the people to remain quietly in their comfortable homes, and warning them that they were liable to contract worse diseases in their huts and camps in the forest to which they were preparing to flee; and that all would in the end regret this hasty and injudicious step. But it was all in vain; I was a comparative stranger and my advice was unheeded. And they all had the dengue, from which they were fleeing, far worse than they could have had it had they remained at home; and as I had predicted they deeply regretted not doing so. Never in its history did yellow fever take such a sweeping course as was exhibited by this epidemic of dengue. And this is one of the important characteristics distinguishing it from yellow fever; for dengue not only takes in cities, towns and villages but the entire region which it has invaded and includes almost the whole population among its victims, seldom, if ever ending fatally as a disease *per se.* Persons who have had previous attacks of dengue, as well as those who have had yellow fever, are equally as liable to dengue as those who have never suffered attacks of either. On the other hand, one attack of yellow fever is protective against a second attack, and the disease, as a rule, is confined to the large cities on the seaboard (though occasionally its germs have been carried in fomities by railroads into interior towns); but its spread has never been of an extensive, sweeping, epidemic character like that of dengue and la grippe.

Another fact in connection with this last epidemic is contradictory of all the past history of yellow fever; namely, that eight out of every ten cases discovered and diagnosed as yellow fever in Galveston, with nearly as many in Houston, all recovered—an unheard of and unprecedented result in that disease. Neither was there any further infection from that number of foci in the two cities, nor could a source of infection be traced in a single one of these cases; but the conclusion of the officials was that somehow it must have slipped in with the dengue. How I wish they had told us from whence came this dengue and also something about its modes of travel and transportation, and by what route it invaded our State !

The foregoing, I think, sufficiently explains my position in the matter; and doubtless ninety per cent of the physicians of Texas agree that I was correct, for subsequent events fully sustained me. Just here let me parallel the principal characteristics and symptoms of the two diseases.

YELLOW FEVER	DENGUE
(1) Is a non-contagious (but highly infectious) epidemic, non-eruptive fever of only one paroxysm lasting from thirty-six to seventy-two hours—a disease of but two stages.	(1) Is a non-contagious but essentially infectious, epidemic, typically eruptive fever of two paroxysms separated by a short remission lasting in all five days (if the second paroxysm is not prevented).
(2) Cause:—A specific, septic, animal poison of known origin and with no incubative period in the person affected.	(2) Cause:—Wholly unknown, but supposed to be some electrical disturbance in the air or earth; closely akin to la grippe.
(3) Onset:—As a rule occurs at night between nine p. m., and three a. m., with no prodromic symptoms except a few hours of exhilaration or stimulation preceding the sudden invasion.	(3) Onset:—Usually occurs in the daytime, from nine a. m. to 3 p. m.; invasion is generally sudden, but this is irregular as is everything else about the disease.
(4) Course:—The most regular of any disease; always exhibiting characteristic phenomena peculiarly its own, variable only in intensity.	(4) Course:—Very erratic; resembling varioloid in thermometric records, yellow fever in march of pulse and scarlatina or measles in eruption
(5) Tongue:—Clean, except a little white furr, and with edges reddened, becoming dry, brown and cracked in the second stage.	(5) Tongue:—Clean but red in first paroxysm; becoming thickly coated and enlarged in second, with offensive breath.
(6) Eyes deeply injected, turgid and yellowish; expression often fierce, anxious and enquiring.	(6) Eyes and physiognomy not peculiar, or simply expressive of the intense pain being endured.
(7) Headache frontal, with intense supraorbital pain; also pain in back and calves.	(7) Headache occipital or parietal, with deep, intense pain back of the eyes; soreness of all muscles; neuralgia in certain nerves and rheumatic pains in all the joints.
(8) The face, neck and upper part of chest flushed, puffy, reddish inclining to purple, yielding later to an orange or bronzed aspect caused by stagnation of the blood in superficial arterial capillaries.	(8) Skin flushed bright red and eyes suffused, but exhibiting none of the other variations of color seen in yellow fever resulting from the sluggish movement of blood in capillaries.
(9) Hemorrhages from the gums, cracked tongue and all mucous surfaces, as well as from kidneys, uterus and stomach.	(9) No hemorrhagic tendency; occasionally obstinate epistaxis and sometimes hematuria.
(10) Urine frequently albuminous, 30 to 40 per cent. Specific gravity low—1.010 to 1.012; suppression common.	(10) Albuminous urine rare; specific gravity high—1.030 to 1.063—sugar more frequent than albumin, but only temporary.

YELLOW FEVER	DENGUE
(11) Muscular prostration very great; convalescence slow and tedious; relapses frequent.	(11) Muscular prostration slight; convalescence rapid; little tendency to relapse.
(12) One attack affords almost certain immunity for the future.	(12) One attack affords no immunity; rather seems to predispose to others.
(13) Mortality very great—25 to 66 2-3 per cent.	(13) Mortality *nil* if uncomplicated and not overmedicated.
(14) A peculiar, putrescent odor exhales from all parts of the body —more pronounced in second stage and in some epidemics than in others.	(14) No odor perceptible except in the last paroxysm when it is from the stomach and bowels only and of the nature of sulphuretted hydrogen.
(15) Liver, stomach, heart and kidneys affected.	(15) These organs show no pathological changes unless they have been previously diseased.
(16) No sequel, except occasionally swellings of lymphatic glands in neck, axilla, groin and testicle.	(16) Frequently followed by boils and carbuncles.

Leaving out the want of parallelism between the pulse and temperature in yellow fever, the foregoing constitutes the principal characteristic symptoms exhibited by the two diseases; but as I have said there is no one single symptom yet discovered that can be said to be truly pathognomonic of either. It is only by grouping the phenomena constituting each disease that the observer is able to form a correct diagnosis; but it soon becomes as easy for the educated and critical observer to diagnose these diseases by this means and to distinguish between them and all other diseases as it is easy for one to recognize and distinguish his friends and intimate acquaintances from strangers by their general personal appearance and make-up.

But I will close by endeavoring to s
it was possible for me to have been in
regard to last year's epidemic; and,
show that my error was simply anotl
tional incontestible proof that I am corr
my positions and predictions in regard
hitherto terrible scourge, yellow fever.
first place I must insist that if the f(
diagnosed as yellow fever, in Galve:
Houston, were indeed and in truth gen
low fever, then all the other thousan
thousands of cases, not only in "the tw
by the sea" but also all those in ever)
town and village of this state as well a:
- of the Mississippi, were bound to be t
_ disease—true yellow fever; it was e
dengue or all yellow fever. There (
no other rational explanation or reas(
why this supposed epidemic of dengue e
a greater resemblance to and possessed
the pathognomonic symptoms of yell(
than it ever did before. And if it is tru
was genuine yellow fever and not dengi
then I repeat what I said in a paper u
subject in *The Alkaloidal Clinic* of Januai
"The scourge of the south has lost it:
nancy; its death-dealing fangs have l
tracted, and it will no longer be dread
has been in the past."
I will now show how this is simply
strong proof of the truth of my theory, ƿ

and predictions as set forth in the preceding pages of this little book. And it should require but little or no argument or explanation to establish the truth of this last assertion. In fact, it is only what would be naturally expected to occur, according to the past history of the disease and the history of what I have attempted to prove was its origin and cause, the African slave trade. And before closing I will mention some very important facts which I have strangely overlooked and omitted to give in their proper order; for they are as strongly corroborative and confirmative of my theory as anything yet adduced.

These facts are as follows : The island of Cuba has two hundred capacious harbors into most of which ships of the line can enter. Slave ships used to enter them, and yellow fever prevailed there as long as that was the case. But their situation and openings into the Gulf were such that nature's great purifier—water—had such free access and egress that as soon as those vile ships ceased to arrive these harbors like those on our Atlantic seaboard were cleansed by the natural action of the waves and tides and yellow fever ceased to prevail. And it is now as unknown in the interior of Cuba as it is in Michigan and Illinois or any of our interior states. This shows conclusively that it did not originate in that island.

I think I have succeeded in proving satisfac-

torily that since the complete suppression and
removal of that cause the disease has also disappeared from and ceased to exist in all parts
of the world where it had prevailed so often and
so virulently during the existence of the nefarious traffic, except from that filthy little pond—
the bay of Havana, Cuba.. I have also shown
that while it has disappeared from all the other
ports of the West Indies, the reason why it has
been so slow in disappearing from this particular harbor is: First, that it has always been the
headquarters for slave ships from the inaugura-
tion of the traffic until its close; and hence has
received into its mud banks a larger material
quantity of the peculiar filth than has any other
port; second, that the sluggishness of its waters
and its narrow outlet makes it almost impossible for those waters to be completely changed
and purified without the help of man, at least in
any reasonable time; third, because it was the
last point on the globe to receive accretions of
that peculiar filth from which the specific poison of the disease is generated—slaves having,
according to the best authentic accounts I can
find, been landed there as late as 1876. From
this it will be seen that no additions of this filth
had been made to its mud banks for twenty-two
years; consequently the slow action of the little
tide by which its waters are affected has in that
time so diluted, attenuated and carried out por-
tions of this filth that it can no longer give rise

to the virulent type of yellow fever of former
days.

Now if this is not a true statement of the condition now existing in the harbor of Havana, and if
the *materies morbi* of yellow fever is still there
in all of its original potency and virulence, then
the prophecies and predictions of the croakers
will have abundant opportunity for fulfilment.
For since the treacherous and villianous blowing
up of our peaceable visitor—the warship Maine
—the stirring up of that filthy mud, the removal
of all quarantine restrictions, and the fleeing of
thousands from that port to our shores (especially at this particular season) will all be the
occasion and means of the introduction of a
greater quantity of the disease-producing germs
of yellow fever than ever came to our shores in
one season before. And if they possess any of
the old-time vigor and virulence of the yellow
fever germs of forty years ago, it will knock my
theory into a cocked hat and scatter my deduc-
tions and historical data to the four winds of
heaven; and in addition to the horrors of a
foreign war we may confidently expect our
country to be scourged from Texas to Maine
and from the Atlantic to the Pacific by this horrible plague.

But I have so much confidence in the historical data which I have furnished in regard to its
cause, and am so well assured that these data
rest so securely upon the firm basis of eternal

truth that they cannot be overturned, that I have no fears of such a result.

It is possible that I am a little premature in my predictions and that the virus in that dirty little pond of Havana is still virulent enough to be capable of producing another terrible epidemic. If so, then the sooner my suggestions are adopted the better it will be for humanity in general.

I propose to cut canals and turn the Gulf Stream into the bay, and thus wash out the last traces of "the most cruel traffic that ever visited any American port from Cape Cod to the Cape of Florida—a traffic that brought in its train a far-reaching Nemesis such as wrong-doing never fails to bring."

FINIS.

www.ingramcontent.com/pod-product-compliance
Lightning Source LLC
Chambersburg PA
CBHW021815190326
41518CB00007B/606